UNDERSTANDING CLIMATE ANXIETY

How should we react to climate anxiety? This accessible book discusses anxiety and other emotions brought on by climate change, examining what climate anxiety is, why it is becoming so prevalent and how it differs from other types of anxiety.

Written by an expert psychologist, the book examines why climate anxiety is developing so rapidly, particularly in younger people. It looks at how it can manifest differently – sometimes as hopelessness or despair, and sometimes as anger which can serve as a catalyst for action. The book dives into the nuance around climate anxiety, questioning what we can do about it or whether climate anxiety should be pathologised at all, given the very real threat of climate change. It considers cognitive biases that underlie information processing and discusses how politics and interest groups affect people's views. Seeking to understand the polarisation that occurs around this topic, the book suggests how we might alleviate climate anxiety without minimising serious concern about climate change.

This highly topical book will be of great interest to students of psychology, environmental science and social science. It will also be of interest to psychologists, mental health professionals and climate communicators, as well as anyone interested in learning more about climate anxiety.

Geoff Beattie is Professor of Psychology at Edge Hill University and Visiting Scholar at the University of Oxford, UK. He is a prize-winning psychologist, author and broadcaster with a PhD in

Psychology from the University of Cambridge. He was awarded the Spearman Medal by the British Psychological Society for 'published psychological research of outstanding merit'. He has published over one hundred academic articles in a range of journals including *Nature*, *Nature Climate Change*, *Environment and Behavior*, and *Semiotica*. Beattie has acted as a consultant to various international organisations with a focus on sustainability, including Unilever, the Leadership Vanguard (established by the CEO of Unilever), and the Born Free Foundation. He is also a member of the International Panel on Behavior Change (IPBC) which aims to collect and integrate knowledge and evidence on environmentally-related behaviour change.

UNDERSTANDING CLIMATE ANXIETY

GEOFF BEATTIE

Routledge
Taylor & Francis Group

LONDON AND NEW YORK

Designed cover image: abracada/Getty Images

First published 2026
by Routledge
4 Park Square, Milton Park, Abingdon, Oxon OX14 4RN

and by Routledge
605 Third Avenue, New York, NY 10158

Routledge is an imprint of the Taylor & Francis Group, an informa business

British Library Cataloguing-in-Publication Data
A catalogue record for this book is available from the British Library

ISBN: 9781032631868 (hbk)
ISBN: 9781032616766 (pbk)
ISBN: 9781032631882 (ebk)

DOI: 10.4324/9781032631882

Typeset in Joanna MT Std
by codeMantra

*I would like to dedicate this book to **Belfast Royal Academy**.
Without the opportunity that great school provided to
this working-class boy, I know that this book and all the others
would never have been written. My life would have taken a
different course. That would have been my destiny.*

CONTENTS

ACKNOWLEDGEMENTS

I am Professor of Psychology at Edge Hill University and Visiting Scholar at the University of Oxford (Wolfson College Oxford Centre for Life-Writing) and owe an enormous debt of gratitude to both institutions for their fantastic encouragement and support. My colleague and co-worker Dr Laura McGuire has been nothing short of immense throughout! My research in the past few years has been supported by Edge Hill University, the Arts and Humanities Research Council and the British Academy and for that I am extremely grateful. I draw upon research I have published in a variety of academic journals, including *Nature Climate Change*, *Environment and Behavior* and *Semiotica*, and in a number of books, which are all highlighted in the text, and I thank my various editors for their assistance along the way. In particular, I would like to thank Professor Marcel Danesi, former editor of *Semiotica*, whose support and intellectual enthusiasm for my work over the years has been invaluable. Lastly, I would like to thank my editor at Routledge, Emilie Coin, and the British Psychological Society for suggesting that I might like to contribute to this book series. It has been my privilege.

WHO AM I AND WHY AM I GIVING YOU THIS ADVICE?

I am Professor of Psychology at Edge Hill University in the UK and Visiting Scholar at the University of Oxford (Wolfson College Oxford Centre for Life-Writing). I am a Fellow of the British Psychological Society, a Life Fellow of the Royal Society of Medicine and a Life Fellow of the Royal Society of Arts. I did my PhD at the University of Cambridge (Trinity College) on the psychology of language and I have won several prizes in psychology for my research including the Spearman Medal awarded by the British Psychological Society for 'published psychological research of outstanding merit'. My books have either won or been shortlisted for a number of literary prizes and have been translated into several foreign languages including Chinese, Taiwanese, Portuguese, Italian, Finnish, German, Romanian, etc. I have published twenty-nine and over one hundred articles in academic journals including *Nature Climate Change*, *Nature*, *Semiotica*, *Autism*, and the *Journal of Experimental Psychology*. I currently have over three-quarters of a million readers on *The Conversation*.

My PhD research in psychology was on thinking and speaking, and involved the analysis of *pauses* in spontaneous speech as evidence of cognitive activity. My mother always thought I was working on '*paws-es*', i.e. the feet of tiny dogs and cats, because she knew that I loved animals. I hated to correct her – that was part of the joy of being from a working-class background. I moved on to analysing what's now called multi-modal communication – the relationship between the various channels of communication including the verbal and nonverbal channels (particularly gesture) – and subsequently began considering persuasion, thinking and decision-making. This led seamlessly (in retrospect) to research on public perceptions of climate change, attitudes and behaviour, and I have now been working in this area for nearly two decades. This research led me to the conclusion that there is a massive gap between what people say about climate change (and sustainability) and what they actually do, and I began to systematically explore this using a variety of techniques and approaches. When it comes to climate change, this gap is likely to provoke high degrees of climate anxiety (after all, who or what can you then trust?) and this is one major thread in the book. I also notice that value–action gap in myself, which is always somewhat disconcerting. I am nothing if not a self-aware psychologist!

Another thread in the book is trauma and ways of dealing with it. Again, I published my first research in the area of trauma, narrative and understanding some twenty-five years ago. This was an area of specific interest to me because of my working-class background in Belfast, growing up during the so-called 'Troubles' in Northern Ireland, that period of civil strife that went on for over two decades and cost many thousands of lives. I watched trauma being created in the streets in front of me. In this book, I consider the existential threat of climate change as a potential major source of trauma for this generation and I try to put it in some sort of psychological context (or rather a variety of contexts, as you will see).

Like many people, I suffer from a degree of climate anxiety. I say 'a degree of' because it does seem to vary from day to day but then again there's so much to worry about in the world. In the book I consider this issue. But I recognise how important climate anxiety is both for the individual concerned and for our common future which depends on joint cooperative action. Climate anxiety can be debilitating, sometimes resulting in eco-paralysis, a loss of hope and a complete failure to act in more sustainable ways. That's another reason, beyond the suffering of the individual, why it's so important for us all.

We know that many things have to change to make things better, but change needs to start with us. I suggest that we need to work more (much more) towards mitigating the effects of climate change and pressurising those in power to do far more in this regard, whilst looking after ourselves in the process. That is critical. I suggest some simple, research-based, psychological techniques to help with this. Simple things that we can all do immediately that might just help.

I hope you enjoy the book. Good luck!

Professor Geoff Beattie, 2 January 2025

Part 1

THE CONTEXT OF CLIMATE ANXIETY

1

INTRODUCTION: WHAT IS CLIMATE ANXIETY?

I start with images on my television screen. Climate change can be seen right now in the mud of Valencia and those buried in it. Many will experience severe trauma. I was reminded of the mud of Flanders in the First World War. A new trauma was born there – shell shock with soldiers stuck in trenches enduring the horrors of war. Those primitive evolutionary responses, fight or flight, were impossible and this shattered the nervous systems of many soldiers. The medical establishment refused to accept this condition as a psychological disorder. They described the symptoms (the tics, the loss of sensation, the mutism) as a 'hysterical' reaction, a sign of weakness or bad breeding, a lack of moral character. The worst sort of victim blaming. They gave the patients electric shocks to force them to make a sound, to 'cure' their mutism. Similarly, climate anxiety is not officially recognised as a psychological disorder and often dismissed as a hysterical reaction, or just plain ridiculous (because some people argue that climate change isn't real). I consider the implications.

'A TERRIBLE AGONY'

It was November 2024, a good time to be writing about climate anxiety. Already in early November, the signs were that 2024 would be

DOI: 10.4324/9781032631882-2

the hottest year on record. It was predicted that the year would end up at least 1.55 degrees centigrade higher than pre-industrial levels, according to the European Copernicus Climate Change Service. The BBC quoted Samantha Burgess, the deputy director of Copernicus, as saying, 'This marks a new milestone in global temperature records.' The symbolic barrier of a rise of 1.5 degrees above pre-industrial levels, agreed by two hundred countries under the Paris climate agreement in 2015, had now been breached. Ed Hawkins, professor of climate science at the University of Reading, was quoted by the BBC as saying, 'The warmer temperatures are making storms more intense, heatwaves hotter and heavy rainfall more extreme, with clearly seen consequences for people all around the world.'

I had spent part of the morning sobbing. Privately, of course. I'm not prone to outbursts of emotion; I keep things bottled up. The English stiff upper lip and all that, even though I'm not English. I'm Irish. It must be an acquired characteristic. I studied at university in England and I work here. I know how to behave.

But it wasn't that figure that had set me off, even such a powerful symbolic figure as 1.55. It was what was on my television.

It was a film about mud. There were endless shots of brown deep sludge, the mud went on forever, waist deep. I knew it was waist deep because there were people in it, and it was up to their waists. They weren't wading through it, that would be the wrong word, it was more like the kind of trudging movement you have to make through deep snow, where every step is a heroic effort, but trudging sounds too clean, too white and crispy, for this. They were filling buckets of the stuff. These were the rescuers from the other side of the town. That's what had made me emotional, just watching hundreds and hundreds of ordinary people come down this road in Valencia with their shovels, brooms and buckets, some were wearing wellington boots, some just their trainers. It was after all the Mediterranean coast in the south-east of Spain. Two hundred people were reported dead so far, but they were still finding bodies, there were bound to be many more in that deep brown sludge. There was drone footage of hundreds of cars overturned and piled on top of each other in the mud, and one survivor said that he knew that many of the cars still had their drivers imprisoned inside. There was an underground car park in a shopping centre full of cars submerged in this thick brown soup. Nobody knew who was in them. The town of Paiporta with a population of 25,000 reported at least sixty-two deaths so far.

The flash floods were like nothing that had been seen before in this region. One handsome Spanish man with a beard, one of the rescuers, was stopped for an interview, he was from the other side of town, the part of Valencia that had been spared. He held it together for a few moments in front of the camera, but no more than that, and then this macho Spanish façade (so beloved of Ernest Hemingway with those images he carried in his mind of the stony-faced brave matador and the bleeding, defeated, gasping bull) suddenly broke apart. 'They died a terrible agony', he said, pulling away from the camera lest anyone see him openly crying. There was little sign of any army personnel helping with rescue, or any official first responders, just local people with shovels and buckets, waist deep in the deluge.

The Mayor of Paiporta, Maribel Albalat, was reported by the BBC as saying, 'When it rains, people normally go down to their garages to get their cars out in case their garage is flooded.' But this rain was described as more like a tsunami and many were trapped in their garages, suffocated by that brown sludge. Six residents of an elderly care home died when the flood water smashed the doors down. They were still on the ground floor. Bodies covered in mud were still being pulled out of this sucking quagmire that the town had become, on the screen in front of me.

The Prime Minister had to go hastily in front of the cameras to promise ten thousand troops would be sent to the area. But it was too late. He said that it was the worst storm in a century in Spain. 'How could anyone have foreseen this?', he asked.

But climate scientists had been warning about the effects of climate change on more extreme weather patterns for decades. The climate and the weather had become one that day. And here it was, not a thousand miles away, not in some location you've never heard of, but in Europe, on our doorstep. I have done enough research on the perceptions of climate change amongst the general public in the UK to know that this is not what people ever imagined in their lifetime (Beattie and McGuire 2018). They thought that climate change might one day show itself on the other side of the world ('probably the South Pacific', they would say, 'or maybe Australia'), not here and not now and not like this. They always thought that it would be something for the future. 'I do worry about my grandkids', they would say, 'but I don't worry about myself', as if this was some noble gesture and not just that temporal bias doing its magic once again, always in the future, never now.

This is what extreme weather looks like. Mud. Bodies pulled out of mud. In a preliminary report, World Weather Attribution (WWA), a group of international scientists who investigate global warming's role in extreme weather events, found that the rainfall which struck south-east Spain was 12% heavier due to climate change. Climate change also made this extreme weather event twice as likely. The event was so devastating because the ground was extremely dry due to the high temperatures and therefore unable to absorb rainwater efficiently.

It was the images, direct and visceral, that I found emotional (never the statistics, even great symbolic figures like higher than 1.5); three images in particular, cumulative in their effects, until I couldn't bear it any longer. The first was the shot of the army of volunteers, just ordinary people, determined faces, trudging through mud, just going to help their neighbours without making a fuss about it. Wave after wave of them. They were the ones that would suffer and be left to fend for themselves. Give them a broom and a bucket.

The second was this shot of a middle-aged woman with long brown curly hair in what remained of her house. She was pulling cups out of the sludge to try to find something that belonged to her mother who had passed away. She was trying to wipe the mud off to see if they were broken. They all were. They were covered in so much mud that you couldn't tell that they were cups until she wiped them. She had used a cloth, now she was using the sleeve of her dress. She had lost everything, the ceiling of her house had collapsed, and the walls had fallen in, all her possessions had gone, she just wanted to find one thing that she could cling onto. The family had lived in that house for three generations, and she was trying to find a single unbroken cup that might connect her to her family's past.

The third image was this young teenager in what might have started off as a white tee shirt shovelling mud into a bucket. Back-breaking work. Of course, her tee shirt was filthy. The interviewer commented, 'That looks like very hard work.' She smiled back at him, 'I'm sweating', so unexpected and so charming. There was a brief pause. Her smile lingered and then dropped. 'Climate change', said her friend in the background. She didn't repeat it or elaborate. It was a look of helpless resignation.

That's what finally broke me that morning.

And then I saw that Donald Trump was on a different channel saying that climate change was a great scam. 'Perhaps the greatest scam the world has ever known.' And that turned my tears, I have to confess, into a sort of unfocused anger.

Science tells us that the climate change represents an existential threat. Science also tells us that we need immediate global action to mitigate the effects of climate change. Many people recognise that we are not doing nearly enough currently and that the future looks extremely bleak. This can give rise to what has been termed 'eco-anxiety' or more usually 'climate anxiety'. The two terms are often used interchangeably in the literature, although eco-anxiety can be used in much broader ways to cover anxiety about broader environmental issues, like deforestation, pollution, loss of biodiversity, loss of animal species including some of our most majestic species because of selfish motives (Beattie 2019), in addition to climate change and its effects. The American Psychological Association (APA) defines 'eco-anxiety' as 'a chronic fear of environmental doom' and describes it in the following terms: 'The chronic fear of environmental cataclysm that comes from observing the seemingly irrevocable impact of climate change and the associated concern for one's future and that of next generations' (APA 2021).

Climate anxiety is conceptualised and measured in a variety of ways in different studies. Some have focused on the negative emotions associated with climate change like sadness or distress (e.g. Eisenman et al. 2015: Searle and Gow 2010). Others have broadened the concept to include cognitions, emotions and behaviours – persistent worrying (cognition), distress (emotion) and sleep difficulties (behaviour) which can all significantly impair an individual's ability to engage fully in work, school or relationships (Clayton and Karazsia 2020). Some have included this criterion of 'functional impairment' in their operational definition of climate anxiety. Schwartz et al. (2022) highlighted and tested the two main subscales of the Climate Change Anxiety Scale of Clayton and Karazsia (2020), which they had called 'cognitive emotional impairment' (e.g. 'I go away by myself and think about why I feel this way about climate change') and 'functional impairment' (e.g. 'My concerns about climate change interfere with my ability to get work or school assignments done'). They found that their most distressed participants emphasised five distinctive themes in response to several open-ended questions: (1) the deadly threat posed by climate change, (2) its immediacy and irreversibility, (3) the anticipated global chaos as a result of climate change, (4) their helplessness in the face of collective inaction, and (5) their behavioural paralysis.

But what is interesting is that many millions of people across the globe, including many very senior and influential politicians, don't believe any of this. They don't believe in man-made anthropogenic climate change; they claim that it's a 'Chinese hoax' or 'fake news'. And they don't believe that there's a legitimate psychological condition called 'climate anxiety'. They view it as sign of hysteria, a hysterical reaction to propaganda pumped out by the media. Climate anxiety is a highly contested term, more than contested, it's attacked, derided, ridiculed. In a book entitled Understanding Climate Anxiety I need to understand how and why it's contested in this way and so vigorously (and, one must add, so emotionally) by some.

Neither climate anxiety nor eco-anxiety is included in the Diagnostic and Statistical Manual (DSM-5) of the American Psychiatric

Association – the standard classification of mental disorders used by mental health professions in the United States (and throughout the globe), revised as recently as September 2024. DSM-5 is used as the authoritative guide to the diagnosis of mental disorders and contains descriptions, symptoms and other criteria for diagnosing mental disorders. The fact that climate anxiety does not feature in DSM-5 means that it is not officially recognised as a mental disorder. Many have argued that the exclusion of climate anxiety from DSM-5 is a good thing. In her powerful book *Generation Dread*, Britt Wray (2022) says, 'many mental health professionals say it is important that it remains excluded. After all, the last thing we want is to pathologize this moral emotion, which stems from an accurate understanding of the severity of our planetary health crisis' (Wray 2022: 21). This makes enormous sense to me. Just imagine what certain sections of the media might do when told that Greta Thunberg, suffering as she does with climate anxiety, has a diagnosable mental disorder (in addition to her diagnosis of Autism Spectrum Disorder).

But if it is not officially recognised will people take it seriously enough? Will they just dismiss the individuals concerned as 'snowflakes', an insult for someone perceived as too sensitive and too easily hurt by the hard realities of life, but who think of themselves as special. I do remember that 1996 film *Fight Club* where those who want to take part in fight club are told, 'You are not special, you are not a beautiful and unique snowflake.' The term stuck. And if it's not in DSM-5 will different researchers define and operationalise *climate anxiety* differently, creating uncertainty about its precise meaning and about its prevalence?

There is a clear and obvious dilemma about whether it should or should not be included. Van Valkengoed wrote in 2023, 'Climate anxiety is not a mental health problem. But we should still treat it as one.' A somewhat confusing title (at first sight) in which she is trying to resolve this dilemma. She points out that climate anxiety is often associated with climate action such as reducing one's carbon footprint or engaging in collective sustainability-oriented actions; indeed it can be a major source of motivation for engaging in climate action (e.g. Whitmarsh et al. 2022). Playing the devil's advocate, she then extends this to its possible logical conclusion – 'This characterization implies that if the goal is to motivate climate action, then scientists should explore ways to sustain or even increase people's climate anxiety' (van Valkengoed 2023: 385). But she clearly recognises that what we as individuals can do to resolve the climate crisis by ourselves in the short term is necessarily limited and that may leave their climate anxiety unchecked. So, this course of action could be perceived (accurately one might add) as both unethical and damaging to the individuals concerned. She continues: 'There is an unresolved tension between medicalizing or pathologizing an essentially normal human response to a crisis and being able to offer appropriate care to people in distress' (van Valkengoed 2023: 385).

She recognises that this tension is not unique to climate anxiety, and she cites *prolonged grief disorder* as a case in point (and highly relevant because grief can be a major part of climate anxiety: see Baudon and Jachens 2021: 14). *Prolonged grief disorder* is defined as 'intense grief lasting longer than a year' and is included in DSM-5 as a form of mental disorder, but again not without significant controversy. Grief (even prolonged grief) is a natural human response to certain types of events, like bereavement or loss, but now we have pathologised it. However, its inclusion can have significant benefits for sufferers in that they can, as a result, access appropriate psychological care for a recognised clinical disorder (if they feel they need it) and receive medical health insurance cover to support the treatment. So clinicians have to choose between two 'evils' when making this choice as to whether to formally diagnose the grieving individual or leave it unchecked (Eisma 2023: 948).

The same argument applies to climate anxiety. Van Valkengoed concludes:

> Being in a constant state of distress is neither desirable nor realistic, and finding ways to reduce climate anxiety should therefore be an important goal for scientists and practitioners alike. While

climate anxiety is not a mental health problem, it is urgent and necessary that we start treating it like one.

(van Valkengoed 2023: 387)

This seems like a thoughtful middle-of-the-road sort of conclusion.

Climate anxiety can be a debilitating psychological condition regardless of its origin, capable of causing functional impairments. Linking a serious mental condition to accurate appraisal of a real cause (like climate change) might be considered to be a very powerful societal message about climate change and its effects. This explicitly identifies the culprit of so much psychological distress. And if it was a recognised disorder, then there would be more systematic scientific research on how to alleviate it. Van Valkengoed points out that currently there is very limited research on interventions to help sufferers cope better with climate anxiety and no randomised control trials to date. In other words, at the moment we are a little in the dark about what exactly we can do to help, based on good sound scientific research. Inclusion in DSM-5 would certainly encourage research on how to alleviate it and, given that feelings of hopelessness are often a core aspect of climate anxiety, it could perhaps help sufferers deal more effectively with their 'eco-paralysis', their inability to act (Innocenti et al. 2023). It could help sufferers use some of the negative emotions associated with climate anxiety (like anger) to drive (and motivate) them to more effective climate action. In addition, it might help some people take climate change more seriously and that would be extremely beneficial in and of itself.

It is clear when we discuss climate anxiety, we do need to constantly remind ourselves that there are adaptive and maladaptive forms of this anxiety (Taylor 2020). In Taylor's words,

Adaptive anxiety can motivate climate activism, such as efforts to reduce one's carbon footprint. Maladaptive anxiety can take the form of anxious passivity, where the person feels anxious but incapable of addressing the problem of climate change and may take the form of an anxiety disorder triggered or exacerbated by climatic stressors.

(Taylor 2020)

There is practical anxiety and paralysing anxiety, motivating anxiety and overwhelming anxiety. In terms of specific climatic stressors, Taylor highlights extreme weather events (like those seen in Valencia in 2024), as well as things like forced migration due to changing local weather conditions as a result of climate change, and then there is all the worry and stress associated with the scientific predictions about our changing planet. These stressors are more diffuse but can be extremely powerful.

There are arguments for and against including climate anxiety in subsequent editions of DSM-5. I can understand both sides and I may be accused of sitting on the fence here (and sitting on the fence is always very uncomfortable for me). Climate anxiety *can* be a great motivator for climate action, but it depends on the level of anxiety, and whether it is manageable or not (Balaskas 2024: 839). The problem is that often it is not; it can be extremely debilitating. So, you can be left with a very serious condition only experienced by 'snowflakes' (in the eyes of a significant proportion of the population), attacked and ridiculed for their experience. Although I did notice coincidentally that many of those with their buckets and shovels in the mud of Valencia were from that exact same snowflake generation. I just felt I had to mention this.

PRE-TRAUMATIC STRESS DISORDER?

But sometimes serious psychological conditions do take a significant amount of time to get recognised by inclusion in the *Diagnostic and Statistical Manual*, especially those emerging from new and unusual conditions. I feel that I have been here before in a sense. Over twenty-five years ago my PhD student Vicky Lee started doing research on trauma with individuals who had experienced a range of emotionally traumatic events. We encouraged them to talk about their fractured and painful traumatic experiences, and in their stories, they would start to integrate their thoughts and feelings, making sense of what had happened to them and how it had happened (Lee and Beattie 1998). We discovered that this narrative construction would often make these intense emotions more manageable and often instigate a sense of resolution which would result in less rumination so that they could finally have a good night's sleep. We became interested in the details of the narratives about the trauma experiences and found that some particular ways of building the story gave them a sense of predictability and control (Lee and Beattie 2000). The research highlighted to us the power of language (in both talk and writing) in dealing with emotionally challenging experiences (Beattie and Doherty 1995). It also highlighted the malevolent power of silence, of inhibition, of socially constructed silence or just plain personal embarrassment or shame in preventing any kind of therapeutic healing.

A number of the individuals who took part in our research had what is likely to be diagnosed as post-traumatic stress disorder (or PTSD):

> a condition which can develop following exposure to an extremely stressful situation or series of events outside the range of human experience, which may manifest itself in recurrent nightmares or intrusive vivid memories and flashbacks of the traumatic event, and in withdrawal, sleep disturbance, and other symptoms associated with prolonged stress or anxiety.

This definition comes from the Oxford English Dictionary. This is a widely recognised condition these days, with the term 'post-traumatic stress disorder' having fairly common usage (with one occurrence every three million words in written English according to the OED). But post-traumatic stress disorder only appeared for the first time in 1980 in the Diagnostic and Statistical Manual, although post-traumatic stress syndrome had been about since the 1960s. The definition of PTSD, as 'a condition which can develop following exposure to an extremely stressful situation or series of events *outside the range of human experience*', might well get us thinking about climate anxiety. What happened in Valencia might well give rise to significant levels of PTSD in those *directly affected*, a clinically recognised anxiety disorder directly linked to climate change and its effects. The concept of *climate anxiety* generally would need to be a broad term covering perhaps a range of distinguishable conditions, including PTSD.

But what about all the others not themselves stuck in the mud of Valencia, but (like me) just nice and comfy at home watching it on television but imagining an uncertain and frightening future. A world full of torrential rain, a world full of mud. How might we describe our negative emotional response? Is that just a form of neurotic behaviour, a hysterical response? Is that the kind of person I am? Neurotic? Hysterical? And what about those people who can vividly imagine what those symbolic numbers like 1.55 degrees centigrade actually mean for humanity, who dwell on it, who can't stop thinking about it?

One student I spoke to recently said, 'I can't breathe when I think about the fact that we've now gone beyond 1.5, that figure has dominated my adolescence. My imagination is now running wild.' My student also told me that she feels she can't mention it to her family or her friends. They get annoyed at her, they tell her that there's worse things to worry about. She says that she suffers in silence. So, can we be traumatised by our imagining of future events?

The term *pre-traumatic stress disorder* in the context of climate change is widely credited as being first used by the forensic psychiatrist Lise Van Sustersen (see Kaplan 2020) as a 'before-the-fact version of classic PTSD'. It arose from observations of climate scientists' responses to climate change (Grose 2020). What they had been predicting for years was now happening right in front of them, changes in temperature, rising sea levels, more intense and unpredictable weather, more droughts, more intense storms. These were now becoming commonplace. And the climate scientists knew that things were only going to get worse, and they were beginning to report feelings of anger, distress, helplessness and depression. You know things are getting serious when the scientists start to break down.

But *pre-traumatic stress reactions*, as a concept, had been used previously by Berntsen and Rubin (2015). They wrote,

> From its introduction in 1980 until 2013, PTSD was classified as an anxiety disorder (American Psychiatric Association, 1980, 2000, 2013), and so concern for future negative events is to be expected. Indeed, apprehension of catastrophic or socially embarrassing future events is often observed in anxiety disorders … Here we show that pre-traumatic stress reactions, in terms of intrusive images and dreams about negative future events, accompanied by attempts at avoidance and increased levels of arousal, are a real aspect of the phenomenology of PTSD, which hitherto has been overlooked. The term pre-traumatic stress reactions here designate disturbing future-oriented cognitions and imaginations as measured in terms of a direct temporal reversal of the conceptualizations of past-directed cognitions in the PTSD diagnosis.
>
> (Berntsen and Rubin 2015: 663)

Berntsen and Rubin (2015) devised a new clinical measure to assess pre-traumatic stress reactions (Pretraumatic Stress Reactions Checklist or PreCL), which used many of the same items found in DSM-5 for PTSD, but now temporally reversed with a focus on the future rather than the past (e.g. 'Repeated, disturbing dreams of a possible future stressful experience?' 'Avoiding imaginings, thoughts or feelings related to a possible future stressful experience?'). They tested Danish soldiers before their deployment in Afghanistan with questions about involuntary images, flash forwards, hypervigilance, nightmares, etc. (all very similar to PTSD but now all with a focus on future events). They found that 'involuntary intrusive images and thoughts of possible future events, their associated attempts at avoidance, and increased levels of arousal were experienced *at the same level* as post-traumatic stress reactions to past events before and during deployment' (my italics). Their overall conclusion was that 'pretraumatic and post-traumatic stress reactions are two subjectively different manifestations of the same underlying phenomenon'.

They also found that those soldiers who experienced higher levels of pre-traumatic stress before deployment had an increased risk of PTSD after their return from the war zone. Their hypervigilance primed their nervous system to react more strongly when anything untoward occurred.

This research would suggest that we need to take stress reactions to future anticipated events very seriously indeed.

'LIKE DISTANT THUNDER'

Post-traumatic stress disorder, of course, is most common in war. In the First World War (1914–18), the predecessor of PTSD, 'shell shock', was first identified. The OED defines shell shock as 'A disorder identified in soldiers in the First World War attributed to exposure to shellfire and characterised by severe anxiety and other psychological disturbances, often accompanied by somatic symptoms such as rapid heartbeat and nervous tic. Now chiefly historical.' Of course, it's now chiefly historical because it has been superseded by the well-respected and well-recognised diagnosis of PTSD.

The early usages of shell shock are interesting. The first use of the term is credited as occurring in the *British Medical Journal* in 1915:

> Only one case of shell shock has come under my observation. A Belgian officer was the victim. A shell burst near him without inflicting any physical injury. He presented practically complete loss of sensation in the lower extremities and much loss of sensation.

This report implies that the 'disorder' is quite uncommon, the front-line doctor had, after all, only seen one case. *The Times* in 1916 wrote: 'A young soldier … who had been rendered dumb from shell shock in France recovered his speech when listening to the humorous song "Any old iron".' So, here you have a soldier who presents as being completely mute, who miraculously recovers his speech when a humorous song is played to him. This is the stuff of miracles or malingerers, according to the esteemed newspaper *The Times*. I think that it's fair to say that *The Times* doesn't believe in miracles. The concept of shell shock had a difficult journey to some sort of acceptance.

I have two points of personal contact with that journey. The first is that I'm a working-class Protestant from Ligoniel, a mill village in Belfast in the North of Ireland. I am a so-called *Ulsterman*, which is often how Protestants from that part of Ireland refer to themselves (Beattie 1992).

My grandfather was in the Ulster Volunteer Force (the Protestant paramilitary organisation formed to fight Home Rule for Ireland) which volunteered *en masse* to serve in the British army on the outbreak of war. He had served in the British army in the Royal Inniskilling Fusiliers in India and in the Boer War in South Africa. After that, things got a bit vague (Beattie 2004). My mother always said that he didn't like to talk about his time in the army. He worked as a rougher in Ewart's linen mill in Ligoniel when he left the army (my mother and Aunt Agnes also both worked there). He died before I was born but I did meet one old neighbour, Isaac the barber, in the 1980s, who had known my grandfather. Isaac was in his nineties when I interviewed him (he had gone down to the docks in Belfast to watch the *Titanic* being launched – he was that old!). Isaac had tried to enlist with his best friend Edward McMurray in the 36th Ulster Division in 1914. Edward was passed as medically fit, but they wouldn't accept Isaac because they explained he had a weakness in one leg. 'I damaged it playing football', he told me.

> I argued with them about it, I begged them to let me join up. I told them that you don't pull a trigger with your leg. But it was no good. I had to stay at home while Edward went off to war. He was a runner – he carried the messages in the 15th Battalion of the Ulster Division.

His battalion fought at the Somme; a battle immortalised in the collective memory of working-class Protestants from Northern Ireland.

The Battle of the Somme on 1 July 1916, the 'Big Push', was the most infamous battle of that most infamous of wars. The bombardment of the German trenches started on Saturday 24 June, a week before the battle itself, and the noise of the guns, it was said, could be heard in London at night, 'like distant thunder'. The plan was for the men to climb up out of their trenches and walk fully loaded through no-man's land. Soldiers were weighed down with a load consisting of a waterproof cape and cardigan rolled on the back of the belt, a pick or shovel, shaving and washing kit, two hand grenades, one in each side pocket, two sandbags tucked into the belt, a wire cutter 'with a white tape tied round the left shoulder strap', 170 rounds of ammunition and a rifle. In addition, 'five flags for indicating the position of the most advanced infantry were to be carried by each company'.

At 1.10 a.m. on Saturday 1 July zero hour was passed on to all the officers in the assembled trenches. Orr (1987) says that some men slept fitfully for short periods of time that night. At 4.00 a.m., dawn broke, and some battalions had an Ulster fry and some strong sugary tea to get them ready for battle. At 6.25 a.m. the final British barrage opened up, and one soldier observed that 'the air hummed like a swarm of a hundred million hornets'. The early mist was clearing, some knelt and prayed. The traditional issue of rum was distributed, but since many men in the 36th were teetotal others got double or triple the amount.

I can imagine it being passed along. I can almost hear the words themselves. This is the test I suppose of whether it's your history or not. 'Yer [never 'your'] man's good living, I'll have his.' I can hear those words in my head. 'Good living', that's what they would have said. 'He's a good livin' fella', they would be saying. 'I'll have his, if he doesn't want it.'

According to Orr some of the men had brought their Orange sashes (the sashes of the Protestant Orange Order) with them and they now started to put them on. I can imagine them getting the folded sashes out of their haversacks, their most precious possession, the one thing that would identify them in that blackened earth, dressing for death, each adjusting his neighbour's sash, as if they were going to the Field (where the Orange Lodges march to on the twelfth of July each year). In the last few minutes before setting off to walk behind a curtain of shells into no man's land, there were short comforting Orange Lodge meetings held. You can imagine a few prayers, said quietly to cover the fear in their voices. There are old black-and-white photographs of the sons of Ulster on their knees praying before the advance.

At 7.10 a.m. the first Ulstermen climbed over the parapet and lay down in long lines, then five minutes later the second wave climbed out and lay down, and then the third, five minutes after that, all waiting for the bombardment to cease. At 7.30 a.m. the whistles of the officers sounded, and the men rose to their feet and started their walk forward.

The war correspondent of The Times wrote, 'When I saw the men emerge through the smoke and form up as if on parade, I could hardly believe my eyes.' As the leading soldiers neared the first German line there were cries of 'No surrender, boys!'

The casualties were staggering but the 36th Ulster Division still managed to break through the first four lines of trenches until they were virtually besieged in such a forward position. 'We got into what must have been at one time a German trench and waited for the rest to catch up with us, but they never came. We could see nothing but Germans all around us.'

Captain Wilfred Spender of the Ulster Division's Headquarters staff was quoted in the newspapers as saying,

> I am not an Ulsterman, but yesterday the 1st July, as I followed the amazing attack of the Ulster Division, I felt that I would rather be an Ulsterman than anything else in the world. My pen cannot describe adequately the hundreds of heroic acts that I witnessed. With shouts of 'Remember the Boyne' and 'No Surrender, boys', they threw themselves at the Germans, and before they could be restrained had penetrated to the enemy fifth line.

The attack was described a year later as 'one of the greatest revelations of human courage and endurance known in history' (MacDonagh

1917). After the war, King George V paid tribute to the Ulster Division, saying 'the men of Ulster have proved how nobly they fight and die'.

The Somme saw slaughter of a magnitude that had never been witnessed before, and pointless in terms of the ground captured. By the time night fell on that first day of the battle, nearly sixty thousand British soldiers had fallen, either killed or wounded, or had been taken prisoner. Twenty thousand had died. 'This was the greatest loss and slaughter sustained in a single day in the whole history of the British army', Winston Churchill later wrote. He added, 'By the evening of July 1, the German 180th Infantry Regiment was again in possession of the whole of its trenches.'

On Tuesday 11 July, when what was left of the 36th Ulster Division left Picardy, they had experienced over 5500 casualties and 2500 killed in two days of the battle. The Ulster Division had lost more than half the men involved in the fighting. Military historians tell us that by mid-November the Allies had advanced five miles at a cost of 450,000 German, 200,000 French and 420,000 British casualties.

I keep a picture on my phone of the Ulster Volunteers marching down Royal Avenue in Belfast in 1915 with their rifles over their shoulders going off to war. The crowds are cheering them on, young boys are up lampposts waving at their older brothers and their fathers. One or two of the soldiers are glancing towards the pavement at their wife, son or daughter calling out to them. One soldier waves back at his family. All the rest have their eyes firmly forward, the proud sons of Ulster marching into the darkness and their destiny (that phrase 'lions led by donkeys' always comes to my mind). I look up old photos of the men from my neighbourhood in history books and find one image of some lads (and they really were lads) paddling in a river with smiles on their faces for the camera, before later that day moving up to the front. The quiet before the storm.

The weather turned bad, and the advance was delayed by a couple of days. It was time to reflect; the mood was sombre. One young soldier wrote to the secretary of his Orange Lodge near Portadown:

> There is no doubt that when you receive this note I shall be dead. There are all the signs that something bigger than has ever taken place before in this war is about to be launched. The more I brood on what may happen the surer I am I shall not survive it. All of us say, 'It'll be the other fellow who'll be killed.' I feel that I am one of those other fellows.

Another wrote, 'The only thing I can compare it with is like waiting for someone to die. You know it's coming, and you wish to God it was over and done with.'

Isaac's best friend Edward fought at the Somme, and Isaac told me when we talked that his friend

> came back without one of his legs and with half a hand missing. He was in a sorry state. He talked about the war a lot, but the funny thing was he never talked about the Somme. He never seemed to want to discuss the Somme itself. He used to sit in his front room, picking bits of shrapnel and bits of dirt out of his good leg when he was reminiscing. The problem with his good leg was that because there were so many little wounds in it, all the dirt of the day seemed to get caught in it. I used to look after it for him.

And Edward was luckier than many of his friends from those little Belfast mill streets of ours. Orr (1987) describes how the Mabins of the Shankill had two boys in the army and the postman had two envelopes for Mrs Mabin but he couldn't face delivering them both, so he held one back until the following day.

I can imagine Mrs Mabin waiting in the scullery day after day, stirring the dishes around the dirty water in the sink, not really cleaning them, just waiting on tenterhooks at the time the postman usually calls, listening out for that loud knock that she knows will make her jump no matter how prepared she is. Then one day there he is, she can see him out the front just standing there on the street, even before he raps the door, and her heart is in her mouth. I can imagine her scullery (like my own run-down mill house in Ligoniel) and its damp, slimy walls, with a fire in the front room and a rusty tin bath on the nail on the yard wall. I can hear her voice. 'Oh Jesus', she'd say. I know how that voice sounds. It's my mother's. 'Oh, Jesus; Jesus, no.'

She makes her way through the front room brushing against the chair. It's a small, cramped room, I can see the furniture, I know where she'd have bought it – in that wee shop on the Shankill. The shop was there for years, generation to generation. She's crying quietly now as she opens the front door. It's summer, there's a flag up outside, it's a Union Jack, and some orange bunting, limp from the rain of the past few days. It's always wet in July.

The postman is holding something out in front of him, he doesn't say anything, his face is just a blur, she reaches out her hand and you can almost see him trying to draw the envelope away at the last minute. Her next-door neighbour is out on the street, she heard the knock, 'Oh Rosemary', she says, 'Oh Jesus, Rosemary, not one of your wee sons', and she rushes out and puts her arm around her neighbour.

The postman doesn't move, and he doesn't say anything. She opens the envelope slowly. She wants to see the name, and for a second the postman feels ashamed. 'It's my Davy, my wee boy', she says. And both women start to sob, but not openly on the street, they go inside for that. Ulstermen and women don't cry in public out on the street where anyone can see them. Her neighbour thanks the postman; she would do that for her friend. And the postman walks off touching the other envelope for Mrs Mabin, to check that it's still there, and he dreads having to come back the next day.

I have thought many times about the men returning without their friends, guilty to be alive, like Edward, haunted with the images of war in the long restless nights, sitting alone and quiet in front of the fire in those cramped mill houses with no privacy, and bursting into tears suddenly but trying to cover it up. It would come out sounding like a guffaw, but it couldn't be that. And their mothers trying to understand, and trying to deal with their own emotions, and starting to get angry at their sons. They wanted things to change but didn't know how.

And then later they'd say to their neighbour, 'He's a changed boy. He was always so happy-go-lucky before, such a happy, wee child before he had the arm off. He was never the same after he had the arm off.' Sometimes she'd address her son directly, 'you'll have to pull yourself together. This isn't doing you any good, you know.' A mixture of fear and confusion and that kind of self-protective anger in those voices. I can see the shoulder of the amputated arm withering over the months and the years, until it is like the shoulder of a child. I can see the sly, unforgiving looks of strangers at the empty sleeve.

Sometimes the returning heroes might talk a little to themselves, reminding themselves of their last conversations with their dead friends. 'I told him that when it was all over, we were going to go down to Bangor for the day and go out on a wee boat and just put our feet up and smoke all day long.' Not great romantic ambitions, just something that I can imagine might have come to them on that morning when they were lying in those great long lines before marching with those heavy packs on their backs through the blackened earth. An image of a rowing boat off Pickie Pool in the one good day of summer somewhere at the front of their mind.

And the mother looking down at her wee boy, his face lit up by the light of the fire flickering on it, old now before his time, sitting flicking his ash into the fire, talking to himself, some ash on his trousers. 'But I never gave him the fag I owed him. Billy died without me ever getting the chance to pay him back.' And the mother not knowing what to do, except get a little angry with her son for being like this. 'You can't sit there for the rest of your life, you know', she would shout from the scullery. 'We've all got to move on.'

Of course, there were some who came home and found that they could talk about what they had experienced to their wives or whoever might listen. I recall a few lines from Michael Longley's poem *Wound* about his father: 'Screaming, "Give 'em one for the Shankill-!"/"Wilder than Gurkhas" were my father's words/Of admiration and bewilderment' (Armitage and Crawford 1998).

It was something to live up to for those of us who came later. Nine members of the 36th Ulster Division were awarded the Victoria Cross for exceptional gallantry. Robert Quigg was also awarded the Medal Order of St George (Fourth Class), the highest honour of the Russian Empire. Those honours cannot be taken away. There were heroes in our working-class streets all right and these became our collective memories. My mother was proud of her father and the men from around our streets. 'They all joined up without any hesitation; they never needed conscription in Norther Ireland', she would say, and their bravery at the Somme was the stuff of legend and my childhood imagination.

'Not a single man turned back', she would say, 'when it was zero hour, when it was their time, they marched forward', and her eyes would mist over.

DEALING WITH THE SILENCE

Yet there is more to the Somme than the statistics and the fleeting heroic stories that were told and retold and became our collective memory. The Somme was more than a few months in France for those who survived, and there was more to it than this for our culture. It seems that many of my grandfather's buddies were not alone in their silence. Many of those who came back from the trenches never talked about the war, and some never spoke. A decade after the end of the war, 36% of the veterans receiving disability pensions from the British government were psychiatric casualties of war.

My mother always said that her father's drinking buddies never spoke about what they'd experienced and one of his pals would just sit and stare in silence at nothing, and she would add, 'he was a lovely wee man, but I was frightened of him when I was a child. He was always blinking. He couldn't stop himself.'

Shell shock was prevalent amongst the troops returning from the war and the commonest symptom of that great disorder of the First World War was mutism. They couldn't say a single word or even make a sound, and this mutism was particularly common amongst the ranks. Shell shock precipitated a crisis in the medical profession in the 1920s because suddenly tens of thousands of men were suffering from forms of 'hysterical' disorders, that is to say disorders with no organic cause, that had previously been thought to afflict only women. Leed (1979) comments that,

> The symptoms of shell shock were precisely the same as those of the most common hysterical disorders of peacetime though they often acquired new and more dramatic names in war: 'the burial-alive neurosis,' 'gas neurosis,' 'soldier's heart,' 'hysterical sympathy with the enemy.' True, what had been predominantly a disease of women before the war became a disease of men in combat.

The question was why?

This brings me to my second personal connection with shell shock. I was professor of psychology at the University of Manchester for eighteen years. A predecessor of mine at the university was Professor Tom Pear, the first full-time professor of psychology in the UK. He was a pioneer of the psychology of shell shock (Pear 1918) and published a classic book on the subject, *Shell Shock and Its Lessons*, with Grafton Smith in 1917. They argued that it wasn't just the 'routine' horrors of war that were responsible for shell shock, but the peculiar psychological conditions of trench warfare – namely the unpredictability of an unseen enemy shelling from a great distance and the fact that any kind of 'fight or flight' response wasn't possible. They wrote, 'One natural way is forbidden him in which he might give vent to his pent-up emotion, by rushing out and charging the enemy. He is thus attacked from within and without.' It was the passivity of trench warfare, as much as anything else, in combination with intense fear and anxiety, that was the root cause of the disturbance.

Leed (1979), reviewing the evidence much later, commented that,

> The most significant variable in the incidence of neurosis was not the character of the soldier but the character of the war. When the war again became a war of movement with the German offensive

of 1918, even though the fighting was intense and exhausting, the incidence of war neurosis dropped dramatically.

But whilst at Manchester, I had the opportunity to view some of the correspondence between Pear and Smith and the medical establishment. Pear and Smith were arguing for the importance of psychological factors in the development of shell shock, that otherwise normal people would break down under sufficient and sustained stress of a particular kind. The medical establishment vehemently disagreed. These psychological explanations were considered dangerous, subversive even. How was one to distinguish shell shock, a hysterical disorder with no identifiable organic cause, from cowardice or shirking?

There was a vitriolic exchange in the pages of the science journal *Nature* between Pear and Elliot, on one side, and Sir Robert Armstrong-Jones on the other. Armstrong-Jones represented the medical establishment, he was the first consulting physician in mental diseases to London Command, and after the war in 1921 was appointed one of the Lord Chancellor's visitors in lunacy, an office he held until 1931. He reviewed their book in *Nature*. In science we sometimes talk of scientists contesting their conflicting views, but here the nature of the conflict is not so metaphorical. This was a life and death struggle for the souls of the soldiers. Armstrong-Jones maintained that, contrary to the views of Elliot and Pear, shell shock was due to physical causes, particularly 'physical weariness, fatigue, exposure, insomnia, exhaustion, and irregular meals – possibly also on occasion malaria and venereal disease', acting on those with 'a family history of insanity, epilepsy, paralysis, neurasthenia, or parental alcoholism … at any rate … some deeply ingrained defect only curable by extinction of the stock or by its repeated crossing with other more stable stocks' (Armstrong-Jones 1917).

Those mute boys from Belfast and Yorkshire or wherever, who had been stuck in those flooded and rat-infested trenches and shelled night and day for months on end, and then on the sound of a whistle had to climb up a ladder into no-man's land and walk steadily forward into machine gun fire, had some essential defect in their genetic make-up (and possibly VD to boot). It was their character that was at fault, their genetically inherited character ('some deeply ingrained defect only curable by extinction of the stock') and their moral character (loose in their sexual habits with VD, implied with no evidence provided). These heroes from our streets, with no man turning back and 'wilder than Gurkhas', were being blamed if they succumbed to shell shock, a new terrible disorder occasioned by a new terrible form of warfare. And not just blamed, but denigrated, with the suggestion that 'extinction of the stock' was the only cure for this condition, implying that more should have died to help solve the problem. This was a new sort of eugenics argument.

The commonest treatment for shell shock, during and immediately after the First World War, was what Leed called 'disciplinary' therapies – you electrocute the patient to cure their mutism. The leading advocate in Britain of such 'medical remedies' was Lewis Yealland, who pioneered the use of electric shock treatment. In his book *Hysterical Disorders of Warfare* published in 1918, Yealland describes in detail his treatment of a typical shell-shock case, a 24-year-old private who had been totally mute for nine months. This soldier had survived the retreat from Mons, the battle of the Marne, the battle of the Aisne, and the first and second battles of Ypres, among others. Having been sent to Salonica to take part in the Gallipoli expedition he collapsed from what he said was the heat and woke up totally mute: 'for five hours he remained unconscious and on waking "shook all over" and could not speak'.

Yealland outlines his particular therapeutic approach:

> in the evening [the soldier] was taken to the electrical room, the blinds drawn, the lights turned out and the doors leading into the room were locked and the keys removed. The only light perceptible was that from the resistance bulbs of the battery. Placing the pad electrode on the lumbar spine and attaching the long pharyngeal electrode, I said to him, 'You will not leave this room

until you are talking as well as you ever did; no, not before.' The mouth was kept open by means of a tongue depressor; a strong faradic current was applied to the posterior wall of the pharynx, and with this stimulus he jumped backwards, detaching the wires from the battery. 'Remember, you must behave as becomes the hero I expect you to be,' I said. 'A man who has gone through so many battles should have better control of himself.' Then I placed him in a position from which he could not release himself, and repeated, 'You must talk before you leave me.'

Electricity was then applied for one hour; at the end of which time the patient could apparently whisper the sound 'ah'. Yealland was obviously pleased with the progress. Yealland reports that he said to the solder: 'Do you realise that there is already an improvement … Small as may seem to you, if you consider rationally for yourself, you will believe me when I tell you that you will be talking before long.' After two hours, the patient tried to get out of the room but was prevented from doing so. The control of the situation resided entirely with the therapist. Yealland writes that he said to the patient, 'When the time comes for more electricity you will be given it, whether you wish it or not.'

VICTIM BLAMING

Yealland was using an extreme form of 'aversion therapy' (really just punishment, I can't say the word 'therapy' here without it sticking in my throat) to change their symptoms, whether they be mutism, tics, or disorders of vision. There was no attempt to understand what the individual had been through, to help them heal or deal with their trauma. The focus was merely on changing behaviour using these electric shocks in this darkened room, to 'cure' the mutism, to get the patient to whisper 'ah', again and again and again, or 'ah', 'bah', 'cah' over and over, on Yealland's signal. For the patient, the nightmares of helplessness in the filthy trenches of the Somme would be replaced with nightmares of helplessness in Yealland's dark treatment room, attached to electrodes and punished again and again. Even when they had been 'cured', Yealland didn't want them to confide in him or attempt to recall what they had been through. He wouldn't shake their hand; he didn't treat them as men. 'You must behave as becomes the hero I expect you to be.' He didn't think that they were heroes, or they wouldn't have succumbed to their trauma like this.

> With recovery of the voice there is usually a marked change in the mental condition of the patient … He will confide in you, and as a rule becomes very demonstrative. At times he will break down emotionally as he refers to his previous hardships; all this, however, can be easily controlled by one who is ordinarily inventive. The handshake must be refused and he must not be allowed to think that a miracle has been performed. The patient is not sent back to his ward until he has been cured; all references to his former condition are discouraged.
>
> (Yealland 1918: 3)

Young men, some very young indeed, were exposed to trench warfare and battles like the Somme, the likes of which had never been seen before in human conflict. The human nervous system was not adapted to this new form of warfare where fight or flight was impossible, and the psychological outcome was a breakdown in movement, in speech, in human functioning. These young men were then transported back to England and given electric shocks until they made sounds and then given more electric shocks until they said words and were 'cured'. Cured by a therapist who warned them not to reflect on what had happened, who would not shake their hand, who would not allow any emotion to be displayed in this darkened room where this crude barbaric, behavioural science held sway. Unspeakable barbarism.

But there was an alternative to the darkened rooms of Yealland, at the more civilised setting of Craiglockhart Military Hospital near Edinburgh, the hospital described in Pat Barker's novel *Regeneration*, where the therapist was the renowned Dr W. H. R. Rivers. The patients were officers and included, among others, the poets Second Lieutenant Siegfried Sassoon and Wilfred Owen. Sassoon was a decorated soldier, awarded the Military Cross in the war in July 1916 for his 'courage and determination' (he was nicknamed 'Mad Jack' by his men for his near-suicidal exploits), but was appalled by what he had experienced. He took a stand against the war and refused to return to duty after a period of convalescence for a bout of gastric fever at Somerville College Oxford. Rather than court-martial him, the Under-Secretary of State for War decided that he was unfit for service and sent him to Craiglockhart. Rivers pioneered at Craiglockhart the talking cure, with, it must be said, articulate and sensitive young officers who were now for the very first time given the opportunity to talk about their experiences in the war. Officers who had previously been told that it was their duty to forget such experiences and had, as befits officers of this social class, in Rivers' own words, done their 'utmost in this direction' (Rivers 1918). Sassoon wrote his famous poem *Survivors* whilst at Craiglockhart in 1917: 'Their dreams that drip with murder; and they'll be proud/Of glorious war that shatter'd all their pride/Men who went out to battle, grim and glad/Children, with eyes that hate you, broken and mad.'

Rivers details how he talked some of these young officers through their experiences. For example, one young officer who was buried after a shell exploded near him had symptoms including severe headache, vomiting, and problems urinating. This officer had apparently remained on duty for a further two months after this particular experience before collapsing altogether after 'he had gone out to seek a fellow officer and had found his body blown into pieces, with head and limbs lying separated from the trunk'. This officer's nights were haunted by dreams of his dead mutilated friend. Sometimes the friend appeared in his dreams as he had seen him in the field that day, but sometimes he appeared with his limbs and his features eaten away by leprosy. The young officer dreaded going to sleep at night, the expert medical advice that he received at the time was that he should merely attempt to banish all thoughts of his young friend from his mind. This medical advice exacerbated the problem because in Rivers' words 'there is no question but that he was striving by day to dispel memories only to bring them upon him with redoubled force and horror when he slept'.

Rivers encouraged the young officer to talk about this deeply traumatic experience, he acted as the guide, pointing out to him (as horrible as it might sound) that the mangled state of his friend's body was conclusive evidence that his friend had been killed outright. This helped the patient enormously, according to Rivers, because he could then provide his own narrative about what had happened to his friend and that he had not suffered like so many of the others. He was no longer having to banish the thoughts and memories of his friend from his mind, and the therapeutic outcome was that for several nights the patient had no dreams at all and then when he did dream about his friend again the dream did not have the same horror attached to it. We would call this 'cognitive restructuring' in modern parlance. This patient, according to Rivers, made a good recovery and was able to talk about what had happened to his friend. Rivers says that for those individuals who have been brooding over their painful thoughts, 'In such cases the greatest relief is afforded by the mere communication of these troubles to another.' This was seen as a very radical form of therapy at the time. And perhaps for the first time the remembering of past events, the effort after meaning, the encoding of images into words was seen as being of therapeutic value rather than something to be avoided at all costs.

When I first learned about shell shock and its victims, it gave me a different understanding of the events at the Somme. The stories that the patients and their therapists provided allowed a different sort of insight into what life in the trenches was really like. There were many brave men who fought and died at the Somme and many brave men, in fact hundreds of thousands more, who had to try to live with the shame of their 'hysterical' response to the unpredictability of death from an unseen enemy during enforced periods of great passivity. Their quite natural reactions to a situation where the normal human responses of fight or flight, which form the very core of our biological being and our evolutionary self, were prevented. And if the soldiers did crack under this pressure, there was little human understanding waiting for them in hospital back in England, unless they were very privileged to begin with. Those who could talk about the Ulster Division being 'wild like Gurkhas' at the Somme were truly the lucky ones.

It took nearly another eighty years before we fully recognised the significance of all of this. There is now a vast amount of evidence in both the psychological and medical literatures, which indicates that traumatic experiences can give rise to severe mental and physical health problems. The American psychologist James Pennebaker put a psychological theory of inhibition forward in 1995, which summarised the processes by which failure to confront traumatic events can result in poorer health. The principal assumption of this theory is that the inhibition of thoughts, feelings, and behaviour is an active process, which involves intense physiological work. Pennebaker reviews the evidence that when the desire to express negative experiences is inhibited over long periods of time, aggregated stress is placed on the body, which results in increased susceptibility to stress-related disease including cancer, high blood pressure, and more physical disease in general. Vicky Lee and I used Pennebaker's theory as the starting point for our own work.

NEW SUFFERING, NEW WORDS

But this is all about an old war, what's that got to do with climate change and climate anxiety? It is relevant because it is an important story of how the medical establishment can fail to recognise emerging forms of trauma created by novel dangerous and unpredictable situations. Human beings had never before in the course of human history had to endure the prolonged horror of trench warfare, unpredictable and deadly, where both fight and flight, the natural human responses, were blocked. Similarly, human beings have never before had to endure this prolonged existential threat from climate change, with its unpredictable nature, striking here, striking there, Valencia today, where tomorrow?

Those who suffer from climate anxiety are regarded by many as neurotic or hysterical, as with shell shock. Climate anxiety like shell shock seems to have a stigma attached to it and I sense that many sufferers don't like to talk about it. A small number, in relative terms, of climate activists, of course, are not silenced in this way. But I think that for many, and we don't have the exact figures, there may be a degree of embarrassment or even shame about admitting it because they know they will face criticism, blame even. Just as soldiers from the First World War suffering from shell shock were blamed for their suffering – bad genes, VD, bad attitudes, bad breeding, 'hysterical like women' is what was said. Traumatised then silenced and then condemned for it is a deadly combination. But there is another unique aspect that we need to consider here, unique to climate anxiety – climate change will affect all of humanity. So why should a person be allowed to claim special 'status' because of something we are all facing? They are derided for experiencing climate anxiety and derided for admitting it, 'claiming to be something special'. Another layer of disapproval. At least soldiers in the trenches experienced unique and individual mental and emotional challenges. They had that going for them (if you pardon my irony).

There are climate activists who very publicly express their fears and anxieties about climate change, but I suspect that they really are the tip of a very large iceberg (and this is one iceberg that is only going to get bigger with global warming). Later in the book in Chapter 10, I will attempt to explore some of the subtle and not so subtle negative attitudes to climate anxiety and those that suffer from it.

In my view, it's clearly useful to bear the history of trauma in mind as we move through the book and into the present times. There may be parallels that are worth bearing in mind. The reluctance of the medical establishment to accept some new form of trauma arising from novel circumstances always needs careful, critical examination back then, and now. It's too easy to blame the marginalised and the voiceless for their own suffering (young working-class boys from Belfast in the trenches; young climate activists trying to get their voices heard and sent to prison for their troubles).

I am not saying that climate anxiety is necessarily like shell shock, and some (I imagine) will be appalled by the comparison, but they do have some commonalities – both 'modern' (in their appropriate historical context) and new forms of trauma, threatening to wider society. Climate anxiety does not necessarily involve some form of post-traumatic stress disorder (or even pre-traumatic stress disorder). But on some occasions, it might well, and there is evidence of localised rises in depression, anxiety, PTSD and substance abuse following natural disasters like floods and wildfires linked to climate change (Morganstein and Ursano 2020; Cianconi et al. 2020). And with these natural disasters, children and young people are particularly vulnerable to the psychological effects (Kennedy-Woodard and Kennedy-Williams 2022). Young people are inherently more vulnerable to additional stress (Wu et al. 2020) and they are also painfully aware that climate change will have a serious effect within their lifetime (51% of people aged 17–34 think that it will, compared with 29% of those aged 55 and over). Taking a more intersectional lens, young people from the Global South (and from marginalised communities in the Global North) are amongst the groups that will be most affected by climate change, partly because of the direct impact of changing weather patterns and partly because of the severe economic consequences of climate change for them, and yet their voices are rarely, if ever, heard in any decision-making bodies related to climate change.

It is not just the direct effects of climate change that we need to consider (like the mud in Valencia). There are the indirect 'secondary' consequences of climate change – the economic consequences, the forced migration, separation, upheaval, loss of home and community, ruptures to one's life. Rural communities in Ghana had to relocate to the capital Accra because their farming practices were no longer viable due to climate change. The migration left the local communities feeling nostalgic, sad and hopeless (Tschakert et al. 2013). Such forced migration, occasioned by climate, has been found in locations in Africa, India, Australia, and across the globe. But there is another major indirect category, which is the psychological impact of hearing about or discovering the wider climate crisis, and our imagination. My student's psychological response to hearing that we had moved beyond that symbolic figure of 1.5 degrees. I remembered what she had told me – 'that figure has dominated my adolescence, my imagination is now running wild'. But as Kennedy-Woodard and Kennedy-Williams (2022) point out, the direct and indirect effects are not always separate; for example, in the Pacific island nation of Tuvalu, the communities experienced the direct effects of climate change, but this was compounded by the additional stress of climate change generally (Gibson et al. 2020).

> There is also the strong argument that as unprecedented weather events become more commonplace around the world, the distinction between being directly or indirectly affected by climate change will become less clear, and examples such as the Tuvalu case study become more widespread.
>
> (Kennedy-Woodard and Kennedy-Williams 2022: 45)

Valencia, of course, has brought this closer to home. Do you think that the trauma of the people of Valencia is not compounded by their worry about climate change more generally? And this will only be added to by the disarray between nations shown at the COP29 meeting in Baku in Azerbaijan in late November 2024.

You might say at this point, what's the problem about clinical recognition? PTSD is already officially recognised in DSM-5. If people suffer from trauma as a result of extreme weather, they can be officially diagnosed. But trauma resulting from climate change is different from the trauma of war. Those soldiers in the First World War returned home. The source of their trauma was in the past (but relived sometimes on a daily basis in flashbacks). With PTSD resulting from extreme weather and climate change, the source of the trauma is still there, in front of them, and may always be there. That's what makes it different, and perhaps deserving of a new label (Climate Change PTSD, part of Climate Anxiety) with additional therapeutic considerations.

The climate crisis didn't reach the esteemed *Oxford English Dictionary* (OED) until 2021 with the inclusion of several 'new' words all linked to climate change, including 'climate catastrophe', 'net zero' and 'eco-anxiety'. It was a big moment. If we are facing a climate catastrophe, it's important that we have the right words to describe it (with official definitions by the most prestigious dictionaries). We do need to be as precise as we can be, when everything ends.

The OED defined 'eco-anxiety' as 'Unease or apprehension about current and future harm to the environment caused by human activity and climate change.' The update to the dictionary was intended to give clarity to the 'new' language around climate change in the run-up to COP26, the 2021 United Nations Climate Change Conference held in Glasgow (but perhaps it should be noted that some of the words around climate change had been around for some time; for example, the term 'climate change' itself was first used in an article in the US magazine *Science, Arts and Manufacturing* on 15 December 1854).

The science editor of the OED, Trish Stewart, was quoted in the *Independent*, 22 October 2021, as saying:

> As world leaders come together to seek solutions to the climate change problem, it has been fascinating, if at times somewhat alarming, to delve deeper into the language we use, both now and in the past, to talk about climate and sustainability. The very real sense of urgency that is now upon us is reflected in our language. What happens next depends on so many factors but, one thing we can be sure of is that our language will continue to evolve and to tell the story.

The OED also introduced 'climate crisis', 'climate emergency', 'climate sceptic', 'climate denier' and 'climate denialism' at the same time. The American Psychological Association (APA), as we have seen, defined 'eco-anxiety' as 'a chronic fear of environmental doom' and described it in the following terms: 'The chronic fear of environmental cataclysm that comes from observing the seemingly irrevocable impact of climate change and the associated concern for one's future and that of next generations.'

Climate anxiety is monumentally important, and a major challenge for society and for psychology (and a whole range of other academic disciplines). The APA survey released in February 2020 found that 56% of US adults said that climate change is the most important issue facing the world today. More than two-thirds (68%) of the adults the APA surveyed reported that they suffered from a degree of eco-anxiety. Nearly half (48%) of the young adults aged 18–34 said that they felt stress over climate change in their daily lives. The APA warned:

> Humans have evolved to adjust to some environmental stressors, researchers have found, through allostasis, the system by which the body responds to stress. The fear and uncertainty you feel during your first major storm in childhood, for instance, doesn't usually stay with you for life. But if those storms keep happening, and growing in intensity, those anxieties are harder to put to rest. The greater the stressors and the longer we are exposed to them, the more likely our responses are to fail – and those stress responses may remain elevated for the rest of our lives.
>
> (American Psychological Association 2021)

These effects will be widespread but are likely to be particularly damaging for indigenous peoples around the globe, who watch as their homelands are destroyed by the changing climate. The environmental philosopher Glenn Albrecht coined the term 'solastalgia' in 2005, to describe 'the homesickness you have when you are still at home', the land you once knew is disappearing or has gone. I am sure this term applies to the residents of Valencia one week on from their tragedy, who now hardly recognise their only home. And, of course, marginalised communities around the globe who have contributed the least to the drivers of climate change will be the ones who suffer most, economically, physically, and mentally, from its effects (and with the least access to appropriate psychological resources). Again, I am reminded of the marginalised working-class communities of Belfast and other cities in the United Kingdom marching off to war in 1914, a war hardly of their making. There they faced new terrors, some couldn't cope and were denigrated on their return, and electrocuted until they made a sound. That was the only 'psychological treatment' for these 'shirkers' (unlike some officers who were treated with respect and understanding). History, if we are not careful, may repeat itself.

The APA warns its members that 'Clinicians need to be mindful that this alarming historical trend of environmental injustice continues today' and that 'Clinicians must be aware that we have patients who are disproportionately affected and need our support.' My image here is of those ordinary people stuck in that deep mud in Valencia.

This is why understanding climate anxiety is so important, so crucial for us all. But let me warn you at the outset, it may at times be an uncomfortable journey. The *Oxford English Dictionary* reminds me that the earliest known use of the term 'eco-anxiety' was in the *Sitka Daily Sentinel* in Alaska in 1990. It was not used in a positive way in its first appearance, but as something of a slur. The relevant sentence in the newspaper reads, 'The biggest winners are well-paid evangelists of the 'eco-anxiety industry'. Whenever, they preach a new 'crisis', their groups' contributions go up' (26 July 1990). There have always been climate change deniers (it's just 'fake news') and climate change doubters (who can't make up their mind), and people who think that 'climate anxiety' (or eco-anxiety) is just a made-up condition or a sign of hysteria or neurosis, and that climate change is a concept used cynically to attack the status quo, US interests, big corporations, the capitalist state, capitalism itself ('it's all just a Chinese Hoax'). In that first use of eco-anxiety, they refer to it as an 'industry', preaching to the rest of us ('well-paid evangelists'), 'well-paid' implying that they're hypocrites, out for no good. I need to consider the minds of climate change deniers in this book and investigate how views can diverge so dramatically. Climate anxiety cannot be properly understood in isolation, but in opposition. It has its opponents, and I want to delve into their psychology and how they think. I will explore in detail the mind of President Donald Trump later in the book.

I have mentioned the definitions of eco-anxiety from the OED and the APA, but there are other definitions from reputable dictionaries. The Collins Dictionary defines 'climate anxiety' as 'a state of distress caused by concern about climate change'. The Cambridge Dictionary, on the other hand, defines it as 'a condition in which someone feels frightened or very worried about climate change'. I quote from several different sources here, not because I like endless lists, but to demonstrate one important thing. These definitions are all different in subtle and not so subtle ways and that means that when it comes to measuring the effects of 'eco-anxiety' or 'climate anxiety' and operationally defining the concepts, we may see significant divergence in perceived severity, prevalence and impact. Depending upon how the condition or state is defined we may think about different causal factors and developmental pathways. We need to bear this in mind. The OED defines eco-anxiety in terms of 'unease' and 'apprehension', neither of which seems to directly correspond to 'anxiety' in everyday usage; they seem less severe and are not themselves equivalent. According to the OED, 'apprehension' means that you 'worry or fear that something unpleasant will happen' whereas 'unease' covers the same ground but additionally it can mean that 'you are not sure that what you are doing is right' (signalled by the word 'or' in the definition). This could be an important addition as this issue concerns many people confronted with the climate crisis. Are they doing the right thing? Are they doing enough? This may give rise to more intense psychological pressure. The Collins dictionary introduces the term 'state of distress' where 'distress' means 'extreme anxiety'; the APA uses the term 'chronic fear' with its emphasis on persistence in the use of the word chronic. State of distress, on the other hand, doesn't emphasise persistence, states by definition are temporary. The Cambridge dictionary introduces 'very worried about climate change' compared with 'worried' as part of apprehension in the OED's definition. Fear and worry are different, worry is more cognitive (it's about thinking), fear is an emotion (like anxiety) with a physiological component (heart beating faster, palpitations, butterflies in the stomach). Some of the leading anxiety theorists view worry just as one part of anxiety, which potentially helps the sufferer deal with the anxiety, as we shall see.

Fear and anxiety are both felt. In the book, I will address the issue of feelings about climate change (and not just thoughts) at length because I think they are critical. Human beings are creatures who both think and feel and how these two systems interact might will be extremely important in terms of how we process information about climate change, how we act or fail to act to mitigate the effects of climate change, how we feel optimistic or pessimistic about possible solutions to the climate crisis, and why we might think that things are hopeless and feel anxious and depressed.

FURTHER READING

American Psychological Association (2021). Addressing climate change concerns in practice. https://www.apa.org/monitor/2021/03/ce-climate-change. Accessed 19 November 2024.

Berntsen, D. and Rubin, D. C. (2015). Pre-traumatic stress reactions in soldiers deployed to Afghanistan. *Clinical Psychological Science* 3: 663–674.

Clayton, S. and Karazsia, B. T. (2020). Development and validation of a measure of climate change anxiety. *Journal of Environmental Psychology* 69: 101434.

Kennedy-Woodard, M. and Kennedy-Williams, P. (2022). *Turn the Tide on Climate Anxiety: Sustainable Action for Your Mental Health and the Planet*. London: Jessica Kingsley Publishers.

van Valkengoed, A. M. (2023). Climate anxiety is not a mental health problem. But we should still treat it as one. *Bulletin of the Atomic Scientists* 79: 385–387.

Wray, B. (2022). *Generation Dread: Finding Purpose in an Age of Climate Crisis*. Canada: Penguin.

Yealland, L. R. (1918). *Hysterical Disorders of Warfare*. London: Macmillan.

2

SHOULDN'T WE ALL HAVE CLIMATE ANXIETY?

I changed my mind about the title of this chapter several times, before I decided to be this blunt, this provocative. It's important at the outset of this book to outline the scientific evidence about climate change to see whether climate anxiety should be considered a *rational* or *irrational* state. After all, if the climate isn't changing or if it's not that serious, then climate anxiety must be an irrational fear, like the phobia that some people have of small innocent spiders. But if the science is definitive and devastating for humanity, then that's a different matter.

Science tells us that climate change is an existential threat caused by human activities. The Intergovernmental Panel on Climate Change (IPCC) is the international body for assessing the science related to climate change. It was set up in 1988 by the World Meteorological Organization and the United Nations Environment Programme to provide policy makers with regular assessments of what is happening to the climate and its impacts. This panel, comprising hundreds of the world's top scientists, is in a unique position, as it works independently of any government. It is charged with regularly reviewing and critically evaluating the vast body of accumulating scientific evidence around climate change. It was awarded the Nobel Peace Prize in 2007 for its work. It is the authoritative guide on the subject of

DOI: 10.4324/9781032631882-3

climate change; all of their reports are publicly available. It is very informative to read the series of IPCC reports to see how the evidence for the role of human activities in climate change has grown clearer and stronger over the years as the crisis has worsened and the scientific data has accumulated. This is reflected in the language the scientists use in their conclusions. Scientists are always cautious in the conclusions they draw, and the earlier reports use terms like 'the balance of evidence' and probabilistic statements like climate change is 'very likely' to have been caused by greenhouse gas emissions, but the strength of the conclusions has changed dramatically. The latest IPCC report published in 2023 says that 'Human activities ... have unequivocally caused global warming.' No balance of evidence, not highly likely, no probabilistic reasoning, no hedging, it's straight in your face.

The evidence for climate change is *unequivocal*, and actually it has been unequivocal for the past ten years. The language in the earlier reports was different. In 1995, the IPCC said that '*the balance of evidence* suggests a discernible human influence on the global climate' (I have added italics for emphasis). In the 2007 report, the IPCC concluded that 'human activities ... are modifying the concentration of atmospheric constituents ... that absorb or scatter radiant energy ... Most of the observed warming over the last 50 years is *very likely* to have been due to the increase in greenhouse gas emissions.' In the 2013 report, the IPCC concluded that 'warming of the climate system is *unequivocal* and since the 1950s, many of the observed changes are unprecedented over decades to millennia ... It is extremely likely that human influence has been the dominant cause.' In 2015, the IPCC concluded that they are 'now 95 percent certain that humans are the *main cause* of current global warming' (IPCC 2014: v, italics added). In 2023, they released their latest 'synthesis report', summarising the science for policy makers and the public. The report reads:

> Human activities, principally through emissions of greenhouse gases, have unequivocally caused global warming, with global surface temperature reaching 1.1 degrees centigrade above

1850–1900 in 2011–2020. Global greenhouse gas emissions have continued to increase … arising from unsustainable energy use, land use and land-use change, lifestyles and patterns of consumption and production … Human-caused climate change is already affecting many weather and climate extremes in every region across the globe. This has led to widespread adverse impacts and related losses and damages to nature and people. Vulnerable communities who have historically contributed the least to current climate change are disproportionately affected.

(IPCC 2023: 4–5)

The IPCC have also concluded that on the basis of the existing evidence a rise in global temperature will have 'severe and widespread impacts on … substantial species extinctions, large risks to global and regional food security … growing food or working outdoors', as well as producing more extreme fluctuations in weather, including droughts, flooding and storms. The conclusions of the IPCC have been endorsed and supported by over two hundred scientific agencies around the globe, including the principal scientific organisations in each of the G8 countries, like the National Academy of Science in the US and the Royal Society in the UK. This is the accepted science of climate change and, of course, it is very worrying for all of us.

An increasing number of people are witnessing the devastating effects of climate change first-hand, with increased adverse weather conditions such as frequent flooding, stronger hurricanes, longer heatwaves, more tsunamis and periods of drought (IPCC 2014; UK Climate Change Risk Assessment 2017). The World Health Organization (WHO) has warned that with temperatures rising and the increase in rainfall we need to be prepared for more illnesses resulting from climate change, including mosquito-borne infections like malaria, dengue and the Zika virus. The WHO report that 'Climate change already claims tens of thousands of lives a year from diseases, heat and extreme weather', and they say it is 'the greatest threat to global health in the 21st century'. Indeed, the World Economic Forum identified climate change as the top global risk facing humanity, a greater risk than weapons of mass destruction and severe water shortages (Global Risk Report 2016).

The evidence suggests that human beings are the most significant contributor to climate change through energy use, population growth, land use and patterns of consumption (IPCC 2023). Currently, CO_2 emissions from human activity are at their highest ever level and continue to rise. Global CO_2 emissions in 2011 were reported as being '150 times higher than they were in 1850' (World Resource Institute 2014; see also IPCC 2014). Although we cannot undo the damage already done with regards to climate change, we do have the power to adapt our behaviour to ameliorate any future effects.

There has, in fact, been scientific evidence for the role of human activity in producing increased greenhouse gas emissions and climate change for a considerable time. Indeed, as far back as 1896 the Swedish chemist Svante Arrhenius calculated the possible effects of doubling the amount of carbon dioxide on global temperatures. In 1965, President Lyndon B. Johnson's Scientific Advisory Council warned that the constant increase in atmospheric carbon dioxide could 'modify the heat balance of the atmosphere'. The IPCC's brief was to review all of the science, and the conclusions, as we have seen, are crystal clear. Those earlier probabilistic statements (the very stuff of scientific discourse) might have suggested to some a degree of doubt about the origins of climate change, which some people jumped on, including Donald Trump, with their own agenda (Beattie 2023). But many now believe the science behind climate change, the language is no longer even slightly ambiguous, and they understand that it threatens us all. None of us will be immune to its effects and that is extremely scary.

However, many millions of people globally haven't got the message. They seem oblivious to what's happening right in front of them. Greta Thunberg famously told us that 'Our house is on fire'. This message went around the world, she became an icon, but millions carried on regardless. It's not that they haven't heard the message, they've heard it alright (it was impossible not to hear it) but they disregarded it. They didn't believe it. In the US, for example, the Yale Climate Opinion Map for 2023 reveals that only 58% of the American public believed that global warming was caused mostly by human activities (in other words nearly half thought that it wasn't caused by human activities); 54% believed that global warming wouldn't harm them personally. So more than half of the American public thought that they would be okay in the end, so they needn't worry about it. The Yale research also found that 49 million people in the US don't believe that the climate is actually changing.

If you are concerned about climate change, and perhaps have followed the science and expert opinion on this matter, and really understood that we need urgent and radical action to deal with climate change, then these sorts of figures are likely to be extremely worrying. Climate change denial and climate anxiety are inextricably linked. If your house is on fire and you are trying to raise the alarm and rushing around to attempt to put the fire out, and your flatmates are sitting there in the lounge watching Netflix and eating popcorn, them telling you just to sit down and chill is not likely to help with your state of anxiety. Especially, when they light up another cigarette and flick ash all over the (highly inflammable) carpet.

Should those of us who believe in climate change and its consequences all have psychological concern, worry, climate anxiety? Perhaps, the answer must be yes, but the right kind.

The science is clear but many people either don't understand it, avoid it, or don't believe it. Given that we need a concerted global response, this is important, and we need to try to understand why. Climate denial undoubtedly exacerbates climate anxiety. Then there's Donald Trump, one of the world's most influential climate change deniers. He has described climate change as 'one of the greatest scams of all time'. He will need to be discussed in a later chapter in some detail.

FURTHER READING

Beattie, G. (2023). *Doubt: A Psychological Exploration*. London: Routledge.

Global Risk Report (2016). 11th Edition. www3.weforum.org/docs/Media/TheGlobalRisksReport2016.pdf. Accessed 10 July 2016.

IPCC (2023). Synthesis Report. https://www.ipcc.ch/report/sixth-assessment-report-cycle/. Accessed 20 November 2024.

Yale Climate Opinion Maps (2023). https://climatecommunication.yale.edu/visualizations-data/ycom-us/. Accessed 19 November 2024.

3

DO HELPLINES ACTUALLY HELP US WHEN IT COMES TO CLIMATE ANXIETY?

If climate anxiety isn't a clinically recognised disorder, then what is it most similar to? I wanted a straightforward answer and I did what I always do in situations like this – I went to the website of the National Health Service (NHS) in the UK (there are similar health-related websites in other countries). The NHS website is often the first port of call for people needing advice about issues to do with their health. It can often be reassuring and point you in the right direction. But did it point *me* in the right direction?

The NHS website didn't help because 'climate anxiety' isn't mentioned. So I went in search of 'anxiety' instead. I thought that this might be a good alternative start. Anxiety is defined as 'a feeling of worry, nervousness, or unease about something with an uncertain outcome'. The NHS also advises me that

> Everyone has feelings of anxiety at some point in their life. For example, you may feel worried and anxious about sitting an exam or having a medical test or job interview. During times like these, feeling anxious can be perfectly normal.

DOI: 10.4324/9781032631882-4

The NHS has done its job, it has that reassuring tone, normalising anxiety for people like me, who don't like to think of themselves as a 'nervous' or 'anxious' person. 'It's only natural', as my late mother would have said (although she did think that a lot of things were only natural, including some really quite odd things). 'We all get a wee bit anxious from time to time, it will be fine', she would have said. It's funny the way we look for reassuring voices wherever we can find them.

But climate change isn't like an exam or a medical test or an interview, it's not going to be over with soon, and it's not just down to me to pull myself together and get on with it, it's always going to be a problem, and it requires everyone to work together. We can attempt to mitigate its effects, but this necessitates concerted cooperative action.

But the NHS website goes on: 'some people find it hard to control their worries. Their feelings of anxiety are more constant and can often affect their daily lives.' The website explains that we call this 'generalised anxiety disorder (GAD)':

> a long-term condition that causes you to feel anxious about a wide range of situations and issues, rather than one specific event. People with GAD feel anxious most days and often struggle to remember the last time they felt relaxed. As soon as one anxious thought is resolved, another may appear about a different issue.

I was wondering if this might be more relevant to climate anxiety. Climate anxiety endures over time, it's also not just one specific event that makes you feel anxious but a wide range of interconnected events/triggers that may spark peaks of anxiety (announcements of backtracking by governments on renewable energy, government hypocrisy on climate change, public indifference or hostility to climate change views, dismissal of one's thinking by friends, waste, high performance cars racing along the street without a care in the world, that usually does it for me, the list goes on), and as soon as one anxious thought lessens (rather than being 'resolved'), another may appear. It certainly looks at first sight like a close parallel of GAD. Indeed, Schwartz et al. (2022) found that both subscales of a measure of climate anxiety that they developed, which they called 'cognitive emotional impairment' (e.g. 'I go away by myself and think about why I feel this way about climate change') and 'functional impairment' (e.g. 'My concerns about climate change interfere with my ability to get work or school assignments done'), were significantly associated with GAD symptoms.

But then the NHS explains what causes generalised anxiety disorder – they identify several factors – our genes, a history of traumatic experiences like child abuse or bullying, a painful long-term health condition, or a history of drug or alcohol abuse. A set of factors all distinctive in one way – they are all things located within the individual, their genes, their history, their illnesses, their substance abuse (with more than a hint of blame). They are all *internal* (like shell shock a hundred years earlier). The NHS is telling us that generalised anxiety disorder doesn't have an *external* cause, it's us that's the problem. There might be triggers in the outside world, but 'generalised' anxiety by definition is felt about 'a wide range of situations and issues, rather than one specific event', so the reasoning goes, it must be the individual. That's the common element.

I don't find this particularly reassuring when I think about climate anxiety (or myself). Of course, it doesn't mention climate anxiety here, but climate anxiety would seem to be more like this than worrying about an exam where the anxiety disappears as soon as the exam is out the way (and stays in abeyance until another exam comes along). And there are only two types of anxiety mentioned, so you naturally start to think about which one is most similar.

This is the medical establishment helping us with what is called our 'attributional reasoning', trying to answer the *why* question. In psychology, an area called attribution theory explains some of the processes we go through. The NHS is giving us a nudge to think about the causes of generalised anxiety in a particular way. The causes suggested by the NHS are all *internal* (to the individual), *stable* (exerting their influence over a long and protracted period of time) and *global* (affecting different things). These three dimensions have been shown to be important in attributional reasoning. Interestingly, Martin Seligman and his colleagues in 1979, in some extremely influential research, had shown that if you are the kind of person who naturally adopts such a pattern of attribution (internal, stable and global) for negative events then it tends to make you prone to depression because you're assuming that all bad things are down to you, they're always going to be there and will affect everything you do. In terms of therapy, Seligman argued that it helps if you do not immediately internalise the causes of negative events or failure but recognise that events are caused by a combination of internal and external factors or sometimes by external events themselves. You failed the exam because it was really difficult (i.e. an external attribution – something about the exam rather than something about yourself), not because you're not intelligent or bright enough (an internal attribution which is also stable and global because intelligence tends to endure and affect many things other than this examination). It may help, in this particular case, if you remind yourself that many people failed the exam, and you have passed many exams in the past.

So, if you assume that climate anxiety has some similarities to generalised anxiety disorder, then the NHS website might not be helping here at all. It seems to be suggesting that if you suffer from a more general form of anxiety, rather than a very specific anxiety (like exam anxiety), then this is because of internal factors, not because of external factors like how the world is, i.e. such an anxiety-provoking place with climate change and all the rest of it (the wars in the Middle East and in Ukraine, the threat of nuclear war, the COVID pandemic, the strong possibility of future pandemics, etc.). Climate change will be operating over a long time period (stable) with many different aspects all capable of eliciting anxiety in different but connected domains (societal indifference, lack of political will, hostile attitudes from climate change sceptics and so on) but it is *external* rather than internal.

Climate anxiety may thus be a rational and enduring response to this situation, not a personal weakness or illness.

Interestingly, the various NHS websites have much less (i.e. 'nothing') to say about climate anxiety, but I did spot one news item from the Oxford Health NHS Foundation Trust from February 2022 which read, 'Psychiatrist urges families to take climate anxiety seriously', which suggests that they are not currently taking it seriously. It quotes a child and adolescent psychiatrist from the Oxford Health Trust, Dr Catriona Mellor, who was a co-investigator on a major global study, indeed the largest and most international survey of climate anxiety in children and young people. The research had found that 59% of young people (aged 16–25) from ten countries, including the UK, Finland, France, USA, Portugal, Australia, Brazil, India, Philippines and Nigeria, were 'extremely' or 'very' worried about climate change, with 45% reporting that it was having an impact on their ability to go about their daily lives. When they tried to talk about climate change, almost half (48%) reported that other people had ignored or dismissed them; 55% reported feeling a sense of helplessness and 40% said that they felt a sense of grief.

Dr Mellor is quoted as saying,

> Climate anxiety is distressing and upsetting but it is also rational. It is not a mental illness. Anxiety alerts us to danger and normally that means we could find potential solutions. However, faced with such a massive global crisis, such feelings can become overwhelming.

It can also, of course, be overwhelming for many older adults.

The original research paper on which Dr Mellor was a co-author is even more hard-hitting, especially on the implications of the research. Hickman et al. (2021) write:

> Climate change has significant implications for both the health and futures of children and young people, yet they have little power to limit its harm, making them vulnerable

> to increased climate anxiety ... This distress was associated with beliefs about inadequate governmental response and feelings of betrayal ... Subjecting young people to climate anxiety and moral injury can be regarded as cruel, inhuman, degrading or even torturous.

They conclude that it is not the fault of individuals, there are clearly major external factors: the existential threat of climate change itself, inadequate government response, betrayal.

Climate anxiety may be very similar to generalised anxiety disorder but it is clearly different from GAD in some respects, specifically in cause. It's not in your genes or in your history of illness or abuse (although we might well need to consider how they may exacerbate it), the causes and the ultimate cure clearly lie elsewhere, like actually doing something meaningful to mitigate the effects of climate change, but there may be things we can all do, not to inoculate ourselves against this existential threat or dismiss it, but to put ourselves in a better position psychologically to deal with it. But I also suspect that there are many different factors involved in the lived experience of climate anxiety, as well as individual coping mechanisms and government inadequacy. There is, for example, societal belief in climate change and vociferous climate denial which puts pressure on the individual. Then there are the politics of denial and what economic and social factors drive that. And then there's the communication of the science and the misunderstandings that arise, as well as psychological defence mechanisms that affect the processing of the messages. Then there's scientific literacy and ignorance, and failure to engage or talk about climate change and the reasons behind that. There's also the stigmatisation of climate anxiety within society and even within the family. Other people will be important to the experience – those who believe in climate change and those who do not – climate change deniers whose beliefs do not align with the science. Their existence, indeed their prevalence, is undoubtedly a major influence on those who do experience climate anxiety. When others can't or don't understand you, it impacts on how you feel about yourself. You shout, and they still don't hear. It's not surprising that helplessness is such a recurrent and damaging feature of climate anxiety. And then, of course, there's what else is happening in the world and the proliferation of global crises.

FURTHER READING

Hickman, C. et al. (2021). Climate anxiety in children and young people and their beliefs about government responses to climate change: A global survey. *The Lancet Planetary Health* 5: e863–e873.

National Health Service (NHS) website: https://www.nhs.uk/mental-health/

Seligman, M. E. et al. (1979). Depressive attributional style. *Journal of Abnormal Psychology* 88: 242–247.

4

IS CLIMATE ANXIETY
MADE WORSE BY OTHER
GLOBAL CRISES?

We are living in very difficult times. Sometimes, it seems the worst of times. The United Nations Office for Disaster Risk Reduction say, 'At no point in human history have we faced such an array of both familiar and unfamiliar risks, interacting in a hyperconnected, rapidly changing world. Decades-old projections about climate change have come true much sooner than expected.' But how do all of these other global crises like wars, environmental crises, pandemics (the list goes on) affect how I worry about climate change. Is worry like a bucket, can it only hold so much? When some new worry comes in, does the bucket overflow and we forget about climate change? That may be a strange metaphor, but it very indirectly captures my feeling of drowning (my head is in that overflowing bucket in my dreams). I was noticing strange patterns in my own behaviour, skipping on social media directly from one crisis to another, from dead children in Gaza to warnings about climate change, with sometimes cumulative effects, but at other times one would take hold and stay there in my brain, now full of worry. Climate change would be pushed out and then returned to, making the feeling worse. Am I distinctive in some way (or just plain odd), or is that the norm?

DOI: 10.4324/9781032631882-5

I was browsing again. Weirdly it's what I often do when I'm feeling concerned about something and when I'm looking for a distraction, something to take my mind off things – social media and then the news, always in that order. I suspect that I'm not alone (not in the sequencing of my browsing but in the fact that I do it when I'm anxious about something), but it rarely if ever helps. It doesn't even distract me, let alone help to make me feel better.

I came across this headline in *The Guardian* on 7 October 2024: 'World Meteorological Organization says water is the canary in the coalmine of climate change and calls for urgent action.' The article described how rivers were drying up at the highest rate in decades, putting global water supply at risk. The headline among the hundreds of others that morning grabbed my attention, I think perhaps I was becoming slightly desensitised to the usual images of climate change, melting glaciers and coloured graphs showing temperatures increasing over time with future projections in ominous dotted lines going up and up. Or overstressed? Perhaps, I couldn't take any more. These days I tend to skip over these, not deliberately of course, but I read them less carefully. But water, now that's a different kettle of fish. That was my first thought, and I laughed at the idiom. We can't live without water. Water shortage and drought is a particularly emotive message, a powerful set of images.

The article may have used an old clichéd metaphor, the canary in the coalmine, but that also was evocative, colourful even in my mind's eye with that yellow canary in that blackened space (an oddly welcoming image for a second), but with my imagination then projecting me and dozens of anonymous others into that coalmine, all in a tight claustrophobic huddle (dozens because we have to all fit in, anonymous because I don't see their faces). We are all crushed together in the dark, unable to see the light from the top of the mineshaft, with the canary singing away, warning us all, but we can't get out quickly enough. The lift to get the miners back out of the mine is broken, my imagination added that bit. I felt a slight momentary panic to be honest as I read that article, my imagination adding and adding to the original image.

But I was already in a dark and slightly fearful mood that day – there was a lot more going on, a lot more news with terrible images, which I had tried consciously to shut out of my mind. It was the one-year anniversary of the October 7 massacre by Hamas, with a year of indiscriminate retribution in Gaza and latterly Lebanon with countless horrifying images and first-hand accounts. Even the news channels were helping me shield my eyes with the pixelated images of children in Gaza dismembered by Israeli fighter jets. The full horror of our world covered with shuddering pixels.

But that morning, oddly I'd been particularly drawn to an article in *The Atlantic* by Amir Tibon, which was an eye-witness account by someone who lived in Nahal Oz, a kibbutz a mile from the border of the Gaza Strip, before the incursion by Hamas. It was the mundane details at the start of the article that drew me in, which made it harder to ignore. I didn't want any more horror, not immediately anyway. It was a story about domesticity and family life, life on a kibbutz. His wife woke him that morning at 6.30 to say that she'd heard a loud noise, she thought that a mortar might have exploded. He thought that he was dreaming and then realised that he wasn't. They were being invaded. Amir described how his elderly father had rung him when he had heard the news about Hamas breaching the perimeter fence. His elderly parents drove ninety kilometres from Tel Aviv to try to help him, dodging the dead bodies of Israelis and Hamas alike lying in the dirt, and stopping to pick up some traumatised Israeli survivors who in his father's words were 'dressed for a party' (they'd been at the Nova Music Festival). It was a story about families and love, morphing into unspeakable horror seen through ordinary eyes, and that naïveté of 'dressed for a party', the elderly father not familiar with music festivals and modern things. That expression stuck in my mind. But I wouldn't have read the article if it hadn't offered something positive at the beginning, and you know it's a very bad world when massacres can have rays of hope embedded within them that draw you in.

And in that year, there were thousands of such unbearable stories emerging from Israel, from Gaza, from Beirut. It was all overwhelming, it had been such a terrible year, and now the water was running out. I talked to someone I've known all my life who has been through so much in her own life (including stage 3 cancer), and who has always shown herself to be incredibly resilient, and she told me that lately she spent most nights crying in front of the television. Truly desperate times. She said that she felt helpless about climate change, about the war in the Middle East, about the Russian invasion of Ukraine … just helpless. I asked myself, is it any wonder that catastrophic events, including climate change, are having a major impact on mental health globally.

Climate change may be the most important crisis that we all face (and have ever faced) but is it exacerbated by these other developments or pushed aside. Can we only have one existential threat in our mind at one time, as van der Linden (2017) suggests? He writes:

> Unfortunately, many studies have shown that in light of issues such as national security, the economy, health care, and other ecological issues such as water scarcity, global warming generally remains a low priority for most people, consistently occupying the lower ranks of the finite 'pool of worry'.
>
> (van der Linden 2017: 22)

We are facing multiple global crises, war, environmental disasters, pandemics, climate change. How do they all interact? Here we sit, with metaphorical canaries singing in the coalmine about water running out, and all too literal bombs raining down on innocent civilians. Perhaps, the latter far away physically and geographically from most of us, but close enough emotionally and psychologically to impact greatly on us. And with warnings of possible escalation of the conflict in the Middle East and the war in Ukraine (escalation both in terms of the countries that could be involved and in terms of weaponry, including the possible use of 'limited' nuclear weapons), wars may get closer still, and new pandemics may be just around the corner (they always seem to be lurking somewhere, like the stuff of nightmares). We are living in extremely anxious times, perhaps unique in human history – when else have human beings been faced by multiple existential crises like climate change and the wars in Ukraine and the Middle East, which could potentially go nuclear? But will these other crises like the wars in the Middle East and Ukraine which are happening right now blot out (temporarily at least) climate anxiety, eclipsing this fear about the future, like a moon passing in between us and the sun, as van der Linden suggests? And if so, what will be the longer-term implications of this? Is it worse if you temporarily stop worrying about something and then (when the wars eventually end) realise that nothing has changed regarding the climate?

I had read two stories that morning, one planted an image of a canary in a coalmine in my head, the other a whole emotional narrative about ordinary families and war that kept me glued to the page. What had the greater effect on my feelings and thoughts, my fears and anxieties, my plans for discussion and action?

In the journal *Lancet Planetary Health* in 2024, a team of researchers from Hong Kong and the UK led by Lau investigated the cumulative effects of several global crises on the emotions and mental health of young people (aged 18–29 years) in five countries (China, Portugal, South Africa, the USA and the UK). They focused on three global issues – climate change, environmental disaster, namely the radioactive water released into the Pacific Ocean following the Fukushima nuclear accident in Japan, and the wars in Ukraine and the Middle East. They were interested in the individual effects of these different crises, as well as their cumulative impact on our emotions and mental health, focusing on levels of stress, anxiety and depression. They also considered the potential mediating role of media exposure and nature connectedness.

We don't just live in a world faced by multiple crises, we live in a world where we seem psychologically dependent on social media, and where many have lost all their connections with the natural world. Social media has greatly increased the availability and the visibility of global crises. Our disconnect from the natural world may well have inhibited our ability to deal with them. Social media allows updates on global crises to be viewed instantaneously (and, as we all know, it often makes for compulsive viewing). It increases awareness and concern, and has very high immediacy, but with no direct involvement, thus impacting greatly on mental health (Tam et al. 2023). Research has found that high levels of concern for climate change resulting from frequent browsing of social media are associated with poor psychological health (Zhao and Zhou 2020). Research has also shown that an individual's perceived value of nature and their degree of nature connectedness can influence the way they emotionally respond to ecological disasters (Clayton 2020). This might have an important role to play in safeguarding the mental health of individuals living in urbanised spaces.

Van der Linden seems to have been right to some extent at least. The study found the strongest 'emotional engagement' was with the ongoing wars. Climate change was in second place with the radiation leak in third. The study also found that young people in China were most emotionally engaged with the environmental disaster and least emotionally engaged with climate change. Clearly physical proximity, no doubt reflecting psychological proximity, is an important issue, the nuclear leak was much closer to home for young people in China.

The strongest specific emotional responses to the ongoing wars were *concern, sadness,* a feeling of *helplessness, disgust, outrage* and *anger;* for *climate change,* the strongest emotional responses were *concern, sadness, helplessness, disappointment* and *anxiety.* Feeling *helpless* was a common response to each of the global crises across the various countries. Young people from the USA felt helpless at similar levels in response to wars and climate change. In Portugal, young people felt similar levels of helplessness in response to all three crises.

Overall, psychological well-being was lowest in China, which reported the highest levels of anxiety (47% of the sample). Levels of depression in young people were highest in South Africa and then the UK, with potential cases of depression estimated to be 50% in South Africa and 46% in the UK, but the figure for depression was higher than 40% for all countries except the USA.

The study also found that media exposure was highest for wars, followed by climate change, with the environmental disaster of Fukushima the lowest for all countries except China (where it was the highest). Connectedness to nature was lowest for the UK and the USA (and highest for China).

To examine the effect of emotional engagement on psychological well-being and distress, the researchers used structural equation modelling to analyse the complex relationships between these variables. They found that emotional engagement with climate change was able to explain 4.3% and 0.5% of psychological distress and well-being variances, respectively, which further improved to 6.4% and 4.0% with the inclusion of mediators. In other words, emotional engagement with climate change did predict psychological distress and to a lesser extent well-being and the association between emotional engagement with climate change and psychological health was strongly mediated by media exposure. Connectedness to nature independently predicted better psychological health suggesting that this is indeed a powerful mitigating factor.

Greater emotional engagement (including concern) was significantly associated with more time spent browsing for respective social media stories on each of the global disasters and it was found that this often leads to a downward spiral regarding mental health. For example, it was found that for many of the young people (especially those from the USA, South Africa and Portugal), 'Compulsively searching for new developments in the war situations might have occurred as part of problematic news consumption patterns, which when perpetuated by addictive and emotionally immersive news stories can evoke additional symptoms of psychological distress' (Lau et al. 2024: e375).

What this detailed and intricate study tells us is that we are living in unprecedented times regarding global crises and each of these crises is having an impact on the mental health of young people. Young people will be particularly susceptible to these, partly because of the psychological impact of the COVID-19 pandemic and the disruptive effect on their educational arrangements but also because these global crises (climate change, war and the concomitant threat of nuclear war, and environmental disasters) are only going to get worse across time and their life span. And they know this. They are also marginalised in terms of decision-making, adding to feelings of helplessness.

Some of the strongest emotional responses were common across the different global crises. These crises make young people feel *concerned*, *sad* and *helpless*, but climate change has that burning level of *anxiety*, compounded by everything else that is going on. Our use of social media and 24/7 news coverage does not help young people cope with any of these, neither does our disconnect from nature. These two factors make it all so much worse. We are constantly shown things that we can't immediately fix on our own (the annihilation of Gaza, the bombing of Beirut, the ballistic missiles and drones into Israel, the horrifying casualties in Ukraine, wildfires, droughts, famine, mass migration). This makes us all feel helpless, like falling down a mineshaft with nothing to cling on to, and yet others criticise us for expressing our fears. We can hear the canary singing but others can't. It's inexplicable. We are falling and we're not just told to shut up. They criticise us for making a noise. If we suffer from climate anxiety then we may well be stigmatised as neurotic, that's what I was also hearing in casual conversation (I was now looking out for those sorts of things). And that may make climate anxiety particularly damaging.

FURTHER READING

The Guardian, 7 October 2024. World Meteorological Organization says water is the canary in the coalmine of climate change and calls for urgent action.

Lau, S. et al. (2024). Emotional responses and psychological health among young people. *Lancet Planetary Health* 8: e365–377.

United Nations Office for Disaster Risk Reduction (UNDRR). Launch of Global Assessment Report on Disaster Risk Reduction (GAR2019). https://www.undrr.org/

van der Linden, S. (2017). Determinants and measurement of climate change risk perception, worry, and concern. *The Oxford Encyclopaedia of Climate Change Communication*. Oxford: Oxford University Press.

Part 2

CLIMATE ANXIETY AND THE BRAIN

5

WHY CAN'T PEOPLE MAKE UP THEIR MIND ABOUT CLIMATE CHANGE?

To understand climate anxiety, we need to understand a little about how the brain works. I like to make a joke sometimes in presentations about psychology and climate change. I will suddenly say, 'people can't make up their mind about climate change because they don't have a mind'. You can feel the tension spread around the room, a quiet ripple at first, slowly becoming audible. Those present think I'm being overly critical about the public, another pompous stuck-up middle-class academic (they don't know my background) looking down on ordinary people. I like to leave a gap, to feel the growing and justifiable wrath. I then follow it up with the punchline – 'they don't have a mind', I say, 'they have two'. I sound like some kind of bad comedian. It sometimes gets a laugh, sometimes just a sigh of relief. I don't know which I prefer. Human beings use two different systems of thought. One of the systems is unconscious, automatic and hidden and it's critical to the perception of threat. It responds to threat by generating 'feelings' that influence our behaviour. This system underpins the development of anxiety, including climate anxiety. We need to understand how. Some people may have the same rational thoughts about climate change but not have the same feelings. This, I think, is a critical difference.

DOI: 10.4324/9781032631882-7

Although climate change is recognised by science to be an existential threat, there seem to be significant differences across individuals in the perception of this risk (Hine 2023). These range from those with extreme climate anxiety to those who see it as of little concern, overblown by politicians or the media, or worse still, 'fake' news. Concern about climate change seems to be higher in some countries (the UK, Australia, Europe) than in others (such as the USA, China and Russia) (Lee at al. 2015). In 2020, I was invited to give a talk on 'How to avoid a global catastrophe' to a party of distinguished Russian journalists visiting the UK, organised through the British Embassy in Moscow. They knew I was going to talk about the environment and seemed very interested in environmental crises (I think they had Chernobyl in mind), but some sat stony-faced when I talked about climate change (there were thankfully a couple of very interested attendees). At the time, I was surprised by this, but they explained that the Russian media did not report or dwell on climate change in the way we do in the West. It wasn't considered 'a big deal in Russia'. A small reminder that it's dangerous to assume that every country thinks alike about climate change.

Climate change is considered more of a threat in the developing world and the Global South (Leiserowitz 2007) where the effects of extreme weather events are perceived first-hand. For too many in the developed world and the Global North, climate change can be more 'abstract', 'a cumulative problem', 'a statistical concept summarising long-term changes in the variability of the earth's climate'. Human beings have evolved to deal with risks in the here and now, not in the perhaps distant future like this. Many in the developed world see it as 'potentially' serious but not as serious as say terrorism, war, health care and the economy (Lorenzoni and Pidgeon 2006), which seem like more immediate threats. The rain and mud in Valencia, however, may have significant effects on the perception of climate change going forward by those living in the Global North. Although many do understand how serious it is, without witnessing it first-hand.

So why does perception of threat vary so significantly across people and cultures, and differ so markedly from the view of scientists?

To answer this, we need to consider how the human brain processes information, including threat information. The answer may lie there, in the basic architecture of the brain.

THE BRAIN IS A STRANGE COMPUTER

The commonest metaphor used about the brain is that it is like a computer; and when we talk about human information processing, we imagine a computer whirring away, rational, and detached from the world, processing the input to arrive at logical conclusions. But that's not how the brain works. How we feel about something is very powerful and there is strong evidence that how we feel about a particular risk can have a more powerful influence than our knowledge and rational thinking about it (Slovic et al. 2002). Computers don't have these feelings and that's why they're different from us (see, for example, the classic 1972 book by Dreyfus).

Psychologists and neuroscientists have developed 'dual-process' theories of how the brain works (Chaiken and Trope 1999; Kahneman 2011). One process is slow, rational and analytic, a process we share with computers (but in the human case this is open to conscious introspection); the other process is much faster and automatic, more 'experiential', derived from our everyday world of associations and actions and feelings. This process operates below the radar of consciousness. We feel certain things about people, and we don't know why. We instantly like certain people, and have a good feeling about them, but would be hard pushed to explain why. And similarly with people we dislike. 'I just can't help it', we say when challenged. This automatic and unconscious system (based around feelings) is often very quick indeed – we make judgements about whether we like someone or trust them literally within a fraction of a second (Willis and Todorov 2006; see also Beattie 2011a).

So different are these two processes that the Nobel Laureate Daniel Kahneman (2011) has recognised them as two distinct *systems* of thought. System 1 is the fast, automatic and largely unconscious system; System 2 is the slower, more deliberate and conscious system.

Take a very simple example, imagine looking at an angry face – as quickly as you recognise the gender of the person or the colour of the person's hair, you have decoded the facial expression – that person is ANGRY. It's automatic, unconscious and fast. This is typical System 1 thinking. You don't have to think about the fact that you have noted that the eyes seem to glare, and the eyebrows are furrowed, or observed that the jaw and lips are tense, with a narrowing of the lip corners, and the nostrils slightly flared. That would take too long and might well be too late. That is an angry person, perhaps out to do you harm!

A multiplication task, on the other hand, is much slower and more deliberate; it requires conscious effort. Try multiplying 13 by 87 and you will see what I mean. This is 'System 2' thinking. That can be *very* slow System 2 thinking for some people!

In everyday life, System 1 is always active, dealing with many of the routine aspects of everyday life (including the decoding of emotional expression). Kahneman describes System 1 as a 'workaholic' and System 2 as sometimes a bit lazy ('harsh ... but not unfair', according to Kahneman (2011: 46). System 1 often jumps rapidly to conclusions, but System 2 doesn't always check the validity of the conclusions, even when it would be easy to do this. Here is a well-known example of this. Answer the following question:

A bat and ball cost one pound ten pence, the bat costs one pound more than the ball. How much does the ball cost?

I often use this example with my students in a lecture theatre full of eager students. They like being asked such a simple question. They often like to be helpful, and they will shout out their answer very quickly to look good in front of their peers.

'TEN PENCE!' booms out. They look very pleased with themselves. There are just a few students sitting quietly who haven't responded. They look as if they might be thinking. The eager ones look disdainfully at them. 'Not so quick off the mark, eh?'

That answer looks great, at first sight. One pound (bat), ten pence (ball) comes to one pound and ten pence. But the problem is that the bat is not then one pound more than the ball, it's ninety pence more. So, the answer isn't correct. The answer is five pence (and one pound five pence for the bat). And there's nothing unique about my students, 80% of university students at some of the very best universities get the answer wrong.

The problem is that the incorrect answer *feels* right but isn't, and System 2 often endorses what feels right without actually going to the trouble of checking the maths (what the quiet students in the lecture theatre were hopefully doing). System 1 is jumping to conclusions here, with System 2 not checking the validity of the conclusions. You can see why Kahneman describes System 1 as a workaholic and System 2 as lazy.

The two systems work on different principles. System 1 works on the principle of associative activation:

> ideas that have been evoked trigger many other ideas, in a spreading cascade of activity in your brain. The essential feature of this complex set of mental events is its coherence. Each element is connected, and each supports and strengthens the others.
>
> (Kahneman 2011: 51)

Feelings are important in System 1 – over time through learning and individual experience, mental representations of ideas, objects and events become *tagged* with affective associations and these affect-laden tags guide subsequent judgement formation (van der Linden 2017).

I should point out that *affect* is a term that psychologists frequently use in this and other domains. It is whether we feel that something is good (positive affect) or bad (negative affect) and this can be relatively stable. It is different from emotion which refers to more specific transitory states like *anger* or *fear* (Leiserowitz 2005: 1436). According to Paul Slovic (2002), affect is also 'an orienting mechanism' that directs fundamental psychological processes such as attention, memory, and information processing. Affect, in particular, is important in the operation of System 1. System 2 uses more symbolic representations and reasoning like a computer (but slower and sometimes with flaws). When it comes to the perception of threat, including the threat of climate change, these feelings-based affect judgements tend to be more powerful (Loewenstein et al. 2001).

There has been an assumption going back many decades that all you have to do to get people to properly evaluate risk, including the threat of climate change, is to give them more information about the threat – issue reports like the IPCC, present the scientific data, educate them. And there is *some* evidence of a positive association between knowledge about climate change and public concern. But its effects are not necessarily that big. Van der Linden (2015) has estimated that knowledge explains only 10% of the variance in public concern. Shi et al. (2016) say the figure varies between 2% and 18% cross-culturally. These knowledge-based attempts have tried to appeal to System 2, our rational self, to feed it with information, and they've completely forgotten about System 1, our experiential self, guided by feelings and associations, quick and sometimes instantaneous. That's why we need to understand more about how System 1 works (and how to influence it).

To understand how System 1 'thinks' (if ever there was a badly chosen word it is here!), then you need to think about actions and events in the real world. A world full of approach and avoidance, scary animals and friendly animals, people you recognise and people you don't, good foods and poisonous ones, likes and dislikes, positive emotions and negative emotions, and feelings that come from nowhere, like mistrust, disgust, danger and envy. You can demonstrate System 1 in operation. Kahneman outlined several experimental studies in his classic book *Thinking Fast and Slow*. System 1 is the 'thinking fast' in the title.

I ask a colleague to run me through one of Kahneman's experiments, one that might illustrate System 1 in operation. It is up to her to decide on the experiment and the stimuli; I ask her to consult his book and find one that I might not know or might have forgotten about (she knows I forget stuff).

So that's why I'm now sitting in a darkened and modern lecture theatre in a psychology department, in a very safe and comfortable environment, but I'm a little nervous. I don't know what she's going to come up with (I kept thinking of the film *Clockwork Orange* and the deconditioning scene), she can be quite cunning, and she turns on the computer.

'Ready?' she asks.

'Sure', I say.

A word pops up. The word is 'BANANA' in bright yellow lettering and a very large font. That's all, just the one word. But let me make a confession. I love bananas. I eat them every day. I am a runner, and all runners love bananas. It's a running thing. Before a marathon, you will see banana skins all over the place (rather than say chocolate wrappers). If you see a runner eating a banana, you don't say 'why are you eating that?' They would find that question very odd. I don't pause in the supermarket before buying a banana. She knows I love them and there it is, right in front of me. I feel myself smiling. I feel mildly hungry. This is a nice experiment.

And then the next word pops up in close physical proximity to the first. The second word is 'VOMIT', bright yellow and large on the screen. She leaves the two words just sitting there for a second or two, no more. It's an odd sort of start to the experiment. What is she trying to prove with this little trick of hers?

Next some instructions appear on the screen. I have to make a word out of a few incomplete letters and shout the word out as soon as it comes to mind. She's filming me. The letters appear 'S – CK'. I shout out 'SICK' very quickly and very loudly. I am, after all, a very cooperative and keen experimental participant. The lights go on.

It was an extremely short psychological experiment!

'Is that it?' I ask.

She thanks me for taking part (she was getting into her role, the only thing that was missing was a white lab coat) and says that as a reward for taking part in the experiment, she would like to offer me a small gift. She holds a tray covered in a tea cloth in front of her. She tells me that I have to make a choice and pulls the tea cloth away.

There is a banana on one side of the tray and a doughnut on the other. My hand moves automatically and without hesitation towards the doughnut.

'Thank you', I say, pleased with her thoughtfulness, and she bursts out laughing (in a very unprofessional way, I thought, nobody likes to be laughed at).

'Why the doughnut?' she asks.

'Oh, I just fancied a change?'

'Nothing else?'

'Like what?'

'Oh, I don't know. Maybe the experiment?'

'That was hardly an experiment!' I replied with a hint of sarcasm (and irritation) in my voice.

'Maybe, it was enough of an experiment', she said calmly. Did you notice how you filled in the blank to form the word?'

'Of course, I said 'SICK', so what?'

'Let me play a little part of a film I made earlier', she said. 'I asked some other participants to do the word completion task, but I didn't show them "BANANA-VOMIT", I showed them instead "BANANA-CHRISTMAS". How do you think they completed the word in that little task?'

'Same as me. "SICK", it's the commonest word beginning and ending like that' (this was just a guess).

And she played the film back to me twice, just to rub it in, and there it was, participant after participant, all six of them, saying exactly the same thing. 'SACK ... SACK ... SACK ... SACK ... SACK ... SACK.'

And that was the experiment.

'Now, isn't that interesting', she said.

She'd caught me off guard because I thought the first slide was just a warm-up. I didn't realise it was going to be the whole thing. I didn't think she was going to be so cunning about it.

She had just demonstrated how System 1 works. It acts in terms of this pattern of spreading activation through association, from *vomit* to *banana*, activating the concept of sickness to one of my favourite foods (which now, temporarily at least, wasn't). And it happens suddenly, without us thinking about it; it catches us off guard, just like she had done. Crafty!

Our minds automatically assume a causal connection, based on the association, between the two words (banana and vomit), producing within us an affective or emotional response (disgust), and this changes, temporarily at least, the state of our memory so that we are now more likely to recognise and respond to objects and concepts associated with sickness and nausea. We are, as she had just demonstrated, more likely to complete the frame 's-ck' as 'sick' rather than as 'sack' or 'sock', having been unconsciously primed with the paired concepts of 'banana' and 'vomit', all because of this associative 'machine' underpinning System 1 thinking. If you use a Christmas prime like 'BANANA-CHRISTMAS', you are more likely to get 'sack' generated (because of the association with Father Christmas and his toys for children). 'BANANA-RUN' might well prime 'sock' in this experiment.

'BANANA-VOMIT' had produced in me a negative feeling about bananas, a temporary affect-based aversion to bananas. When has that ever happened before? I love bananas, I know I love bananas, I've never been sick eating one, I know that. But this single associative link influenced my behaviour.

Kahneman argues that as human beings we don't necessarily understand the causes and operations of our own cognitions and behaviour because of this fundamental division in our cognitive processes with one system hidden and unconscious. We identify with our own System 2 – rational, thoughtful, careful. I read the IPCC reports, I am a psychologist and a scientist, but I can be thrown off track by a momentary feeling. Kahneman writes:

> When we think of ourselves, we identify with System 2, the conscious, reasoning self that has beliefs, makes choices, and decides what to think about and what to do. Although System 2 believes itself to be where the action is, the automatic System 1 ... is effortlessly originating impressions and feelings that are

the main sources of the explicit beliefs and deliberate choices of System 2.

(Kahneman 2011: 21)

There is a very important point here. Decades of consumer advertising have planted a whole range of subliminal associations in our mind. Big cars have been associated with success, this or that perfume connected to sexual allure, designer goods related to status. The advertising industry has worked on influencing System 1 through the process of unconscious association with remarkable success over many, many years. Marketing cigarettes so effectively for so long, a product that can and does kill you, is one of their most inglorious achievements (see Beattie 2018a). To live more sustainably many people will have to override the positive *affect* of high carbon lifestyles and products implanted by all those years of dedicated advertising. My own belief is that this conflict between our feelings about certain things and our rational thinking is one contributor to climate anxiety. It certainly is for me.

As a child in a slum house in Belfast in the 1960s I collected silver-plated lapel pins of Mercedes and Jaguar cars (my father was a motor mechanic but died when I was young). I wrote to the manufacturers and explained that I was thinking of buying one of their cars. I requested a catalogue and I also asked for a nice shiny silver lapel pin. Although what they thought of a letter with a child's handwriting from Legmore Street as a potential customer is anyone's business. But they sent them to me anyway. And I find myself smiling to this day when I see a Jaguar or Mercedes on the road. I have very positive *affect* about these cars even today and it has been a battle at times to override this, and to stop myself following my System 1 into their respective showrooms to buy one of their cars. And the problem is that I sense that I'm not alone. Many of us may well have these two systems in opposition. I called one of my earlier books *The Conflicted Mind* for a very good reason!

The media also uses the 'banana-vomit' paradigm every day in every article, editorial, comment, alignment of image and text; it works through this process of spreading activation. We are then primed to recognise new information, to fill in the blanks, to jump to conclusions. 'Climate change – hoax' or alternatively 'climate change – global catastrophe', which are played out and repeated in different media (including the Twittersphere of Donald J. Trump), set up different patterns of spreading activation. But these associations are built over a lifetime of experience. These associations in System 1 then influence what we endorse in lazy System 2. We often jump to conclusions without examining the evidence because of the activation of System 1 and its emotional undercurrents. This distinction between two types of cognitive process in everyday cognition may help us understand some of the cognitive biases that we are susceptible to when we think about climate change, and that our *feelings* about climate change may often be of paramount importance (sometimes like my manipulation by my colleague in the lecture theatre with her banana and doughnut experiment, or more systematically by the media or those out to influence how we think and feel about climate change).

FEELINGS AND THOUGHTS

Paul Slovic and his colleagues (2002) have investigated how these feelings lead us to think in a biased way about the world. He calls it the *affect heuristic* where people let their likes and dislikes determine their beliefs. 'Affect' is the word used by Slovic to describe this feeling state either with or without consciousness. Slovic et al. write: '*Affective responses* occur rapidly and automatically – note how quickly you sense the feelings associated with the stimulus words *treasure* or *hate*. We argue that reliance on such feelings can be characterized as the affect heuristic' (2002: 397). This idea had been discussed previously by Zajonc (1980) who had also suggested that affective reactions to stimuli are the quickest and most automatic responses, which guide subsequent information processing. As Slovic says, 'Although analysis is certainly important in some decision-making circumstances, reliance on affect and emotion is a quicker, easier and more efficient way to navigate in a complex, uncertain and sometimes dangerous world.' Emotion then can direct your search for and processing of relevant information.

In domains where there is a lot of information (like climate change) and we don't have time to read it all, your affective response can direct you to focus on certain bits of information rather than others. Your emotional attitudes to exercise (positive!), genetically modified foods (negative!) or climate change mitigation strategies (positive!) affect what you attend to. I read an article on the BBC website about the benefits of exercise for healthy ageing several times yesterday; my crafty colleague (with the banana and doughnut experiment), who coincidentally is younger, seems to have ignored it completely. You might imagine that the conscious and rational System 2 has the 'ability to resist the suggestion of System 1, slow things down, and impose logical analysis. Self-criticism is one of the functions of System 2' (Kahneman 2011: 103). However, in the light of the work on the *affect heuristic*, Kahneman writes:

> System 2 is more of an apologist for the emotions of System 1 than a critic of those emotions – an endorser rather than an enforcer. Its search for information and arguments is mostly constrained to information that is consistent with existing beliefs, not with an intention to examine them. An active, coherence-seeking System 1 suggests solutions to an undemanding System 2.
>
> (2011: 103–104)

Tversky and Kahneman have called this the 'confirmation bias'. We seek information that supports what we already believe and that is concurrent with our feelings about it. Your affective response then determines your beliefs about the benefits and risks. If you like exercise, then you probably believe that the risks are low, and the benefits are very high. If you are presented with the relevant information, then you may process it and recall it in a biased way. There is a degree of 'cherry-picking' of the evidence that supports our underlying beliefs. George Marshall in his book *Don't Even Think About It* says that this operates at all kinds of levels, even in terms of the perception of the weather itself. He writes:

> When asked about recent weather in their own area, people who are already disposed to believe in climate change will tend to say it's been warmer. People who are unconvinced about climate change will say it's been colder. Farmers in Illinois, invited to report their recent experiences of the weather, emphasized or played down extreme events depending on whether or not they accepted climate change.
>
> (2015: 15)

So, the same basic pattern of weather can be used to support one view or the opposite view with a cherry-picking of the information and the (often implicit) comparison data ('colder' or 'warmer' are, after all, comparative terms).

Kahneman's work offers us a coherent framework for thinking about biases and how they might operate when thinking about cognitive biases in any context, including climate change. It emphasises the importance and primacy of our affective responses in determining our rationally held beliefs and offers a theoretical perspective on the very different processes that underpin our everyday cognitions. The associative processes of spreading activation are so significant because they are the processes that are used by the workaholic System 1, that hidden, unconscious system that determines so many of our core beliefs as endorsed (rather than tested) by System 2. We may like to think that we are System 2, that we are that rational, reflective system that thinks carefully about the world, before making decisions, but we would be badly mistaken.

Antonio Damasio has explored experimentally how activation of the affective/feelings/emotional system (System 1) precedes activation of any conceptual or reasoning system (System 2). He first showed this (Bechara et al. 1997) with a very simple gambling experiment. Sitting in front of the participant are four decks of cards, in their hands they have $2000 to gamble with. The task is to turn over one card at a time to win the maximum amount of money; with each card you either win some money or lose some money. In the case of two of the decks, the rewards are great ($100) but so too are the penalties. If you play either of these two decks for any period of time you end up losing money. On the other hand, if you concentrate on selecting cards from the other two decks you get smaller rewards ($50) but also smaller penalties and you end up winning money in the course of the game. But these 'reward/penalty' factors are unknown to the participant at the start of the game.

What Damasio found with people playing this game was that, after encountering a few losses, normal participants generated skin conductance responses (a sign of autonomic arousal) before selecting a card from the 'bad deck' and they also started to avoid the decks associated with bad losses. In other words, they showed a distinct emotional response to the bad decks even before they had a conceptual understanding of the nature of the decks and long before they could explain what was going on. They started to avoid the bad decks on the basis of their emotional response.

In other words, Damasio and his colleagues demonstrated that 'in normal individuals, non-conscious biases guide behaviour before conscious knowledge does. Without the help of such biases, overt knowledge may be insufficient to ensure advantageous behavior.' In normal people, activation of the emotional system precedes activation of the conceptual system, and we now know the neural connection between these two systems is in the ventromedial prefrontal cortex.

Subsequently, Damasio demonstrated the powerful role of emotions in the generation of moral judgements – those patients with bilateral damage to the same brain region, the ventromedial prefrontal cortex, were more likely to choose 'heroic' and highly emotional personally aversive responses in a series of moral dilemmas presented to them. Haidt (2001) developed a new model of moral judgement (and evaluative judgement generally) in which moral judgement (and *evaluative judgement*) appear in consciousness automatically and effortlessly, but that 'Moral reasoning is an effortful process, engaged in after a moral judgment is made, in which a person searches for arguments that will support an already-made judgment.' In other words, we make our mind up quickly and the 'arguments' presented to us may play little role in our judgement except in the subsequent justification of our behaviour to ourselves or others.

This work is directly relevant to climate change. It suggests that our affective response to climate change is of critical importance. If you've experienced the devastating effects of climate change first-hand, your affective response will be very different from the responses of those who have not. You will feel differently about it, and these feelings will affect your thinking and your perception of the threat. Of course, some people may well experience a strong affective response to films about what is happening or even to the scientific data presented in cold, hard figures, but not everyone. If we want more people to act to mitigate the effects of climate change, we need to bear in mind that System 1 is feelings-based and we need to get that involved through messaging that will impact on affect, rather than by providing straightforward information.

My colleague influenced my feelings about bananas in a second or two, but, of course, it didn't last that long. Associations like this are evolutionarily adaptive, making us primed to respond in a certain way before going on to the next challenge in the primitive world of our ancestors. But sometimes associations stick, as we shall see, in both adaptive ways (prime sickness with the word 'vomit', avoid a particular food) and non-adaptive ways (for example, the influence of relentless consumer advertising on non-sustainable behaviour), which may be very relevant to both climate change mitigation and climate anxiety. In the meantime, we must remember that people don't have a mind, they have two.

And this unfortunately is no joke.

FURTHER READING

Beattie, G. (2018). *The Conflicted Mind: And Why Psychology Has Failed to Deal with It*. London: Routledge.

Damasio, A. R. (1994). *Descartes' Error: Emotion, Reason and the Human Brain*. New York: Putnam.

Kahneman, D. (2011). *Thinking Fast and Slow*. London: Penguin.

Marshall, G. (2015). *Don't Even Think About It: Why Our Brains Are Wired to Ignore Climate Change*. London: Bloomsbury.

Slovic, P. et al. (2002). Rational actors or rational fools: Implications of the affect heuristic for behavioral economics. *The Journal of Socio-Economics* 31: 329–342.

van der Linden, S. (2015). The social-psychological determinants of climate change risk perceptions: Towards a comprehensive model. *Journal of Environmental Psychology* 41: 112–124.

6

WHY DON'T WE DO MORE?

I think if we want to understand climate anxiety, we need to consider why we all aren't doing more to mitigate the effects of climate change. People say they really care about the planet, that they *really*, *really* care, but their actions don't match what they say. I think that this more than anything else sparked my own climate anxiety. Which philosopher said, 'Hell is other people'? And then, of course, I noticed this within myself, which is even more worrying.

We are faced with compelling scientific evidence that we are experiencing a climate emergency that requires immediate and concerted action at all levels in society, including the political, economic, social and individual level (IPCC 2023). Social and individual action necessitates significant change in our everyday behaviours and in our decision-making about energy use, transportation and everyday consumer choices. We need to change both our slow, rational and considered decisions (like deciding on a new car or holiday, home or abroad, or deciding whether to insulate the family home), and our more habit-based decisions (like turning off lights when leaving a room or choosing local rather than imported produce in the supermarket – I'm always in a rush, no time to think). However, there is often a major disconnect between what people say about climate change (it's important and we need to do something about it) and their behaviour, which seems more resistant to change and

DOI: 10.4324/9781032631882-8

wedded to high carbon products and high carbon lifestyle behaviour (Beattie and McGuire 2018). This is a major issue for us all.

Gifford (2011) has discussed this discrepancy in terms of the psychological 'dragons of inaction'. These include, he says, feelings of learned helplessness (people have tried in the past and it seemed to make no difference so they've given up trying) and lack of self-efficacy (people don't think that *they* can make any difference), emotions not sufficiently negative to drive behaviour, habits resistant to change, and behaviours not predictable from standard measures of our attitudes (which are often very positive towards sustainability). Dealing with these dragons of inaction is a major societal concern.

Various behaviour change campaigns have been introduced in the past to promote awareness of climate change and to encourage more sustainable behaviours. Some of these campaigns over the past decade or so have been media-based, including television commercials (Act on CO_2), magazine advertisements (WWF) and social media (Climate Coalition), but all have had somewhat limited success (Beattie and McGuire 2018). Other campaigns have tried a different approach. With groceries accounting for, on average, one-third of household CO_2 emissions (Moser 2015), it's important to attempt to influence everyday consumer behaviour towards low carbon, more locally sourced food products. Such a change would then, in theory, drive the market to produce fewer high carbon items (a politically more attractive option than say a carbon tax on groceries).

One approach that has been tried here is 'carbon labelling', which provides information about greenhouse gas (GHG) emissions in CO_2 equivalents that can be ascribed to goods and services. Consumers are informed of the environmental impact of the products through a simple labelling scheme, enabling them to reduce the CO_2 emissions of their household by making simple and relatively small changes to their lifestyle (Beattie and McGuire 2018). This was introduced in the UK by the supermarket chain Tesco in 2007. The Department for Environment, Food & Rural Affairs (DEFRA) and others seemed confident on the basis of *self-report* attitude surveys that the UK public were ready to adapt their patterns of consumption 'with appropriate information about how to act' (DEFRA 2016). These surveys had consistently reported that the public held strong views about climate change and very positive attitudes towards more sustainable consumption: for example, '70% of people agree that if there is no change in the world, we will soon experience a major environmental crisis' and '78% of people say that they are prepared to change their behaviour to help limit climate change' (Downing and Ballantyne 2007). Forum for the Future wrote that '85% of people reported that they wanted more information about the associated environmental impacts of their purchases' (Berry et al. 2008). DEFRA (2008) concluded that 'Many people are willing to do more to limit their environmental impact, they have a much lower level of understanding about what they can do and what would make a difference.'

There was a certain confidence when Tesco, the UK-based retailer, introduced carbon labelling back in 2007 (I was there in the audience at the launch at the Royal Society in London). Tesco aimed to include carbon labels on all of its 70,000 own-brand products (see Figure 6.1). This was to be the start of a great British 'green revolution', led by a supermarket chain, of all people.

Terry Leahy, CEO of Tesco at that time, said, 'The green movement must become a mass movement in green consumption.' It sounded admirable. To achieve this goal, Leahy argued, 'we must empower everyone – not just the enlightened or the affluent'. The carbon labels hit the shops, and Tesco waited with bated breath. So did I.

But it didn't work.

Tesco dropped this plan in 2012, arguing that other supermarkets had not joined them in this enterprise, and said that the accurate calculation of carbon footprint was slower and far more expensive than originally anticipated. However, in reality, carbon labelling didn't work because consumers didn't understand what the figures meant, and more importantly they didn't 'feel' enough about carbon footprint to actually attend to the labels. I conducted some research on this with the backing of Tesco (through the Sustainable Consumption Institute at the University of Manchester set up with their financial

Figure 6.1. An example of the carbon footprint label on a Tesco lightbulb. (from *Beattie* 2010)

backing). The results of the research were not good news for Tesco (Beattie 2012). Using eye-tracking to chart individual gaze fixations when viewing products (with both front and back views of the product presented simultaneously), we found that people paid very little attention to the carbon labels.

We analysed gaze fixations on a variety of Tesco products (Tesco Greener Living light bulb, orange juice, non-bio detergent, etc.) every forty milliseconds and found that in less than 7% of all cases did participants fixate on either the carbon footprint icon or the accompanying carbon footprint information in the first five seconds (Beattie et al. 2010). Five seconds is important because that is the average length of time we view a product before making our choice in a supermarket.

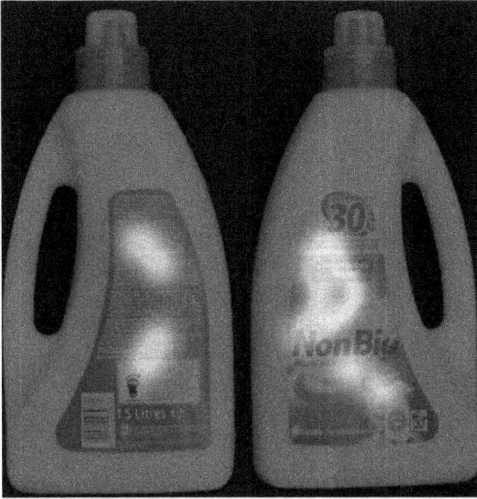

Figure 6.2. Sample fixation hotspots on 'non-bio' liquid detergent. (from *Beattie* 2012)

See Figure 6.2 for a 'hot spot analysis' of one participant viewing an image of a detergent – light colouring shows the pattern of fixation. You will notice that it is not on the carbon footprint.

Our experimental participants were not behaving in the way anticipated by either government or major retailers. They were not noticing or reading the carbon labels. Other research from Manchester University at the time found little understanding of the values attached to carbon labels (Upham et al. 2011), and carbon footprint was ranked 13th out of 14 on the list of important attributes of a product (Gadema and Oglethorpe 2011). But most importantly, unlike the inclusion of nutritional information like fat content and calories on products, which Tesco had recently introduced, and which produced an almost immediate change in some consumer choices, carbon footprint had little effect on consumer behaviour (Beattie 2012). Tesco discontinued carbon labelling in the UK in 2012, although it continued in various forms in other countries, and is now being reintroduced in the UK (although it remains to be seen as to whether these labels will now draw consumers' attention in the 5–7-second time window when choice is made).

Despite expensive carbon labelling and other major initiatives, large-scale behavioural change in the direction of more sustainable lifestyles simply did not occur. And when we consider factors that might impinge on climate anxiety, this level of apathy amongst the general public must be one of the major ones. People are not changing their behaviour despite the climate crisis. For example, domestic energy consumption increased by 40% between 1990 and 2005 and by 1.5% between 2015 and 2016 (Department for Business, Energy and Industry Strategy 2017). One of the world's leading multinationals, Unilever, outlined in their 'Sustainable Living Plan' in 2013 how they planned to halve the GHG emissions of their products across the lifecycle by 2020. They reduced GHG from their manufacturing chain, doubled their use of renewable energy, produced concentrated liquids and powders, reduced GHG from transport and refrigeration and restricted employee travel. The result was that GHG footprint per consumer 'increased by around 5% since 2010' (Unilever 2013). Unilever concluded that the biggest challenge facing this whole enterprise was 'consumer behavior', which was more resistant to change than had been previously anticipated, and was, in fact, moving in the wrong direction (despite the apparent increasing positive attitude of the public to more sustainable lifestyles). There have been some changes in consumer behaviour in the interim but not of the magnitude required.

Some of this resistance to change may be attributable to the public thinking that their behaviours will not make any real difference to the overall issue, they feel disempowered, they have feelings of low self-efficacy (feeling that they personally can't make a difference to climate change mitigation) and low response efficacy (feeling that their behaviours will not make a difference). The resistance to change may also be associated with feelings of learned helplessness where people have 'learned' to stop trying, they have tried in the past to do something and it didn't seem to have any effect (Seligman 1972). For this reason, assessing and attempting to change empowerment

(feelings of self-efficacy and response efficacy) and learned helplessness need to be central features of our efforts. In the final chapter of
this book, I will outline some of my own recent research in this area.

But behavioural inaction may also be due to something else − it
may be attributable to the fact that although people know that climate
change is bad, they don't feel it sufficiently strongly or they believe
that it won't impact on them personally because of low perceived
vulnerability and optimism bias, which we will consider in more
detail in Chapter 7 (see also Beattie et al. 2017). This has been a
major focus of my work in recent years.

EXPLICIT VERSUS IMPLICIT ATTITUDES

There does unfortunately seem to be a big discrepancy between people's expressed attitudes towards climate change and sustainability and
their actual behaviour, between what people say and what they do.
This is sometimes referred to as the 'value-action gap' (Kollmuss and
Agyeman 2002). For example, in one study, although 50% of participants reported that they preferred to buy organic products, this was
contradicted by actual consumption data (Tsakiridou et al. 2008).
Ajzen and Cote (2008) concluded that the predictive validity of explicit
attitude measures declines as social desirability increases. In the case
of climate change and sustainability, the social desirability of certain
responses is very high (Beattie 2010). In domains like sustainability,
self-reported, explicit attitudes may be overshadowed by social desirability and reporting biases (everybody knows what to say to look good,
even most politicians, with some exceptions like Donald Trump).

So why do people not behave in accordance with their (expressed)
attitudes to sustainability? Why don't they do more to help mitigate
the effects of climate change?

In my view, this raises some fundamental psychological questions: for example, as to whether people do have the right attitude
in the first place to sustainability/climate change, and whether they
do just need better information (like carbon labels) to direct their
behaviour?

An attitude was defined in 1935 by Gordon Allport, one of the founders of social psychology (and the 'father' of attitude measurement), as 'a mental and neural state of readiness to act'. Some psychologists have argued that *self-reported* attitudes, which necessarily are both conscious and explicit, may not be sufficient for understanding and predicting behaviour, and have questioned whether 'our mental and neural state of readiness to act' is necessarily conscious in the first place (Beattie 2010; Greenwald et al. 2009). Indeed, Allport himself, who first developed self-report measures in the discipline, seems to have held doubts about this. He wrote: 'Often an *attitude* seemed to have *no representation in consciousness* other than a vague sense of need, or some indefinite or unanalysable feeling of doubt, assent, conviction, effort, or familiarity' (Allport 1935, my italics).

One alternative approach to this issue of the potentially weak relationship between self-report measures of attitudes and actual behaviour, is to measure 'implicit' attitudes, where reporting biases may not be so prevalent. Greenwald et al. (1998) defined implicit attitudes as 'actions or judgments that are under the control of automatically activated evaluation, without the performer's awareness of that causation' (Greenwald et al. 1998: 1464). In other words, based on System 1 processes. Allport displayed some awareness of these in his classic 1935 volume. He wrote:

> The meagreness with which attitudes are represented in consciousness resulted in a tendency to regard them as manifestations of brain activity or of the unconscious mind. The persistence of attitudes which are totally unconscious was demonstrated by Muller and Pilzecker (1900).
>
> (Allport 1935: 801)

He clearly did not rule out the concept of unconscious attitudes but chose to focus exclusively on the measurement of attitudes with self-report questionnaires. I have argued elsewhere (Beattie 2013) that his reasons for this particular focus were not just academic but highly personal (he had visited the famous Sigmund Freud in Vienna when he was still a student, and Freud had tried to psychoanalyse him, it put him right off the idea of the unconscious for good!). Allport's legacy, both academic and personal, then defined attitude measurement in psychology, and related disciplines, for many decades to come.

But interest in 'the meagreness with which attitudes are represented in consciousness', in other words 'implicit cognition' and 'implicit attitudes', has been growing, and this could lead us to think very differently about the 'value-action' gap. This research might one day tell us that the 'value-action' gap does not actually exist because we have been measuring and factoring in the wrong measure of 'value' in the first place. Kahneman, as we have seen, argues that as human beings we do not necessarily understand the causes and operations of our own cognitions and behaviour because of this fundamental division in our cognitive processes.

> When we think of ourselves, we identify with System 2, the conscious, reasoning self that has beliefs, makes choices, and decides what to think about and what to do. Although System 2 believes itself to be where the action is, the automatic System 1 … is effortlessly originating impressions and feelings that are the main sources of the explicit beliefs and deliberate choices of System 2.
>
> (2011: 21)

Greenwald et al. (1998) have considered the accumulated effects of all of this associative activation for attitudes, our 'mental and neural state of readiness', and argued that we may well have implicit attitudes formed on such basic processes that are not available to introspection and are indeed unconscious. The problem with this theorising about implicit attitudes was that we had no way to access implicit attitudes or measure them reliably, until Greenwald developed the Implicit Association Test or IAT (Greenwald et al. 1998). The basic premise is that when participants categorise items into two sets of paired concepts, if the paired concepts (like 'low carbon' and 'good') are strongly associated, participants should be able to categorise the items faster, and with fewer errors, than if they are not strongly associated (see Figure 6.3 for an example of a high/low carbon footprint IAT).

In some domains, consciously reported explicit attitudes and implicit attitudes measured through speed of association are correlated (although the size of the correlation does vary), but in many other domains, there seems to be little or no correlation between the two measures. This seems to be the case in the sustainability domain (Beattie and Sale 2009, 2011), and other 'sensitive' domains like race (Beattie 2012, 2013). This has led Greenwald and Nosek (2008) to suggest that explicit and implicit attitudes can be 'dissociated'.

Implicit attitudes might be particularly relevant to sustainable consumer choices. Our decisions as consumers are *sometimes* slow, deliberate and 'thoughtful', but are, on many occasions, driven by much more automatic and non-deliberate processes (Panzone et al. 2016),

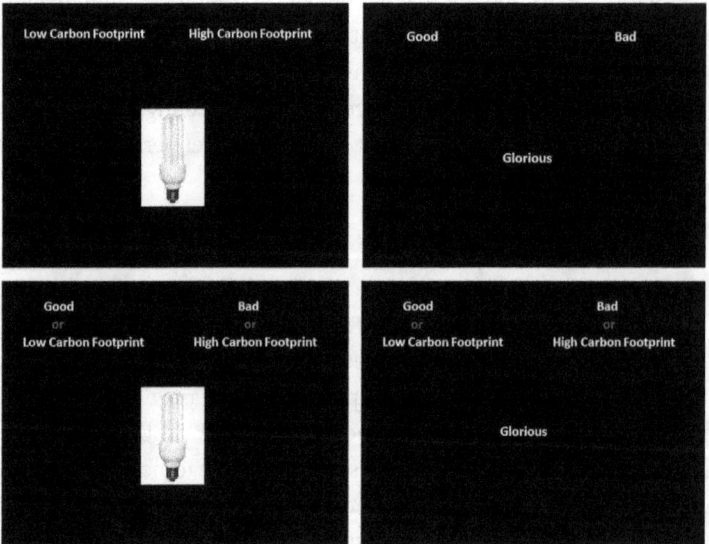

Figure 6.3. An example of a carbon footprint IAT.
(from *Beattie* 2010)

that are core to everyday shopping *habits* (Beattie and McGuire 2014). Panzone et al. (2016) have argued that the automatic evaluations in consumer decision-making are based on the experientially derived implicit attitudes towards various products (Eagly and Chaiken 2007). Bettman et al. (1998) and others have argued that 'consumers often do not have well-defined existing preferences but construct them using a variety of strategies' (1998: 187). One such 'strategy' is the implicit evaluation of the product (Bohner and Dickel 2011), and the operation of implicit processes.

Implicit attitudes are assumed to be based on a slow-learning associative system (my colleague presenting me with 'banana-vomit' would be one tiny fragment of that associative system), whereas explicit attitudes are based on a fast-learning system, which uses higher-level logic and symbolic representations. Rydell and McConnell (2006) have shown that you can change implicit and explicit attitudes with different sorts of information. Consciously accessible verbal information about a target changes the explicit attitude towards that target, whereas subliminally presented primes, 'reflecting the progressive accretion of attitude object-evaluation pairings', change the implicit attitude towards them. You can even change implicit and explicit attitudes in opposite directions by using associative information below the level of conscious awareness to change implicit attitudes, and consciously processed verbal material (in opposition to this) to change explicit attitudes.

So, what are the possible implications of this 'divided self', of not having one but two underlying types of attitudes, for behaviour in general, and more specifically for consumer behaviour in the context of the threat of climate change?

Both types of attitudes can be relevant for behaviour, but under different sets of circumstances and this is what the empirical research seems to suggest. Self-reported attitudes may predict behaviour under certain situations, especially when people have the *motivation* and the *opportunity* to deliberate before making a behavioural choice (Fazio 1990), but they are less good at predicting spontaneous behaviour under time pressure, or when consumers are under any sort of cognitive or emotional load. Unfortunately, time pressure, cognitive load and the absence of any opportunity to deliberate characterise much of everyday supermarket shopping. Supermarket shopping is rarely found to be a slow, deliberate, reflective process, the shopper passes about 300 brands per minute and each individual choice is often quick and automatic. In such contexts, unconsciously held implicit attitudes might be a better predictor of actual consumer behaviour than explicit attitudes, where an implicit attitude is defined as 'the introspectively unidentified ... trace of past experience that mediates R' (where R is the response – the favourable or unfavourable feeling, thought, or action towards the social object) (Greenwald and Banaji 1995: 5).

In other words, habitual consumer behaviour without much opportunity or motivation to deliberate might be driven by processes not available to introspection and therefore not picked up by self-report measures. They require a different sort of measure. Greenwald and Banaji (1995: 5) wrote: 'Investigations of implicit cognition require indirect measures, which neither inform the subject of what is being assessed nor request self-report concerning it.'

The concept of implicit attitude gives us a different way of thinking about the motivational basis for human action and could be a critical element in the fight against climate change. Implicit, rather than explicit attitudes may well be underpinning everyday habitual consumer behaviours. Such behaviours may be 'sticky', in sociological jargon, because attempts to change attitudes and behaviour just focus on certain types of messages, ignoring the associative networks of the implicit system.

IMPLICIT ATTITUDES AND BEHAVIOUR

Over the past decade or so, my colleague Laura McGuire and I have been investigating how implicit attitudes relate to how we process information relevant to climate change, assuming that the processing of relevant information is the start point of the whole process of behavioural change.

In one study, we analysed the relationship between explicit and implicit attitudes and visual fixation on carbon labels on products (Beattie and McGuire 2015). We found that there was no significant relationship between explicit or implicit attitude to carbon footprint in terms of the overall amount of attention devoted to the carbon label. However, there was a significant statistical association between our measure of implicit attitude and the target of the first fixation. Those individuals with the most positive implicit attitude were more likely to fixate first on the carbon footprint information (rather than say 'energy' or 'price') compared with those with a more negative implicit attitude. Those with the most positive implicit attitude had a mean of 7.0 first fixations on carbon footprint whereas those with the least positive implicit attitude had a mean of 4.5 first fixations on carbon footprint. This association did not, however, occur with explicit attitude. Those with the most positive explicit scores had a mean of 5.3 first fixations on carbon footprint whereas those with more negative explicit attitudes had a mean of 6.5 first fixations on carbon footprint. This difference was both in the wrong direction and non-significant. So again, we find evidence that measures of implicit attitude, but not measures of explicit attitude, predict patterns of unconscious eye movements.

This research opens up the possibility that we may have implicit attitudes at odds with what we report (and indeed at odds with how we think about ourselves), which can nevertheless influence our everyday behaviour. But the question remains, to what extent do these implicit attitudes predict consumer choice? In a very simple study, Beattie and Sale (2011) had found that when participants were asked to select either a high carbon or low carbon goody bag at the end of an experiment measuring attitudes, those with a strong pro-low carbon implicit attitude were more likely to select the low carbon goody bag, but only under time pressure. Very similar results had been reported by Friese et al. (2006) who also found that implicit attitude predicted the choice of a gift (a 'generic' gift versus a 'branded' gift) for taking part in the experiment, but again only under time pressure.

These results are interesting, but of course tell us very little about how people will behave in a real consumer choice situation because of broad ecological considerations. Consumer products are characterised by a number of different dimensions (brand, value, taste, health features, environmental implications, etc.), all operating simultaneously, which could impact on consumer choice at both the associative and more rational levels. Advertising is used to build brands (be they well-known brands, luxury brands, organic or eco brands, or value brands) in an associative manner (Aaker and Biel 2013), and when it comes to consumer choice under time pressure, even when System 1 might be more active, these other associations might swamp any implicit associations to do with our attitudes to carbon footprint.

These considerations formed the basis for another study (Beattie and McGuire 2016), where we analysed consumer choice of real brands as a function of time and as a function of both implicit and explicit attitudes. In this study, we found that consumers are very sensitive to both brand information and value in their selection of products. The brands chosen most frequently under no time pressure were the well-known brands – Heinz, Kellogg's, Hovis, etc. (chosen in 38.0% of all selections, with four alternatives to choose from) – followed by the value brand (32.4%). Significantly further down the list were the organic/eco brands with 17.0% and lastly the luxury brands at 12.6%. When behavioural choice was made under time pressure, this trend became even more pronounced, and the well-known brands were selected even more frequently (pointing to the power of advertising for promoting brand recognition). Well-known brands were now chosen in 42.8% of all cases and value brands 31.4% of the time (down slightly). Organic/eco brands were now in last place with only 10.4% of selections.

This time dimension (so characteristic of much supermarket shopping) had a statistically significant effect on consumer choice in terms of the selection of well-known brands compared to organic/ eco brands. Under time pressure, consumers were significantly more likely to choose luxury brands and significantly less likely to choose organic/eco brands. Given the social and temporal aspects of much supermarket shopping, often characterised by significant time pressure, this is not an optimistic conclusion regarding environmentally sensitive choices.

We also found that those participants in our study with a positive implicit attitude to carbon footprint were guided by colour-coded carbon footprints but not by the numerical values of carbon footprint, representing the gradations of high and low carbon, within them. Given that most countries haven't introduced colour-coded carbon footprints but have instead opted for numerical values on a plain background, this might well explain why these campaigns have up to now been relatively unsuccessful in promoting behavioural change and selection of the low carbon alternatives. We also found that under time pressure, the strong pro-low carbon implicit group did show a significant tendency in selecting low carbon items; the weak pro-low carbon implicit group did not show a significant tendency in this regard.

In other words, when participants/consumers are under time pressure (as they are in many everyday consumer situations), those with a strong implicit attitude to low carbon are more likely to shop in a sustainable way. Our measure of explicit attitude to low carbon also significantly predicted the choice of organic/eco products, but here only when the choice was not made under time pressure, suggesting that they may need more time to process the label and/or reflect on the nature of their choice.

The results of this study gave us some insight into the variables that affect consumer choice and help point towards the attitudinal measures that may allow us to predict more sustainable consumer behaviour. In the case of implicit attitudes measured using the IAT, one might say that it is extraordinary that a simple reaction time measure, which computes the response time in a categorisation task, can predict anything at all in a separate domain. However, this measure predicts choice of low carbon items and even predicts the choice of organic/eco products (at least when there is more time for the consumer to reflect). The advantage of this simple measure is that participants do not seem quite so able to distort it for reasons of social desirability, in order to appear greener than they really are,

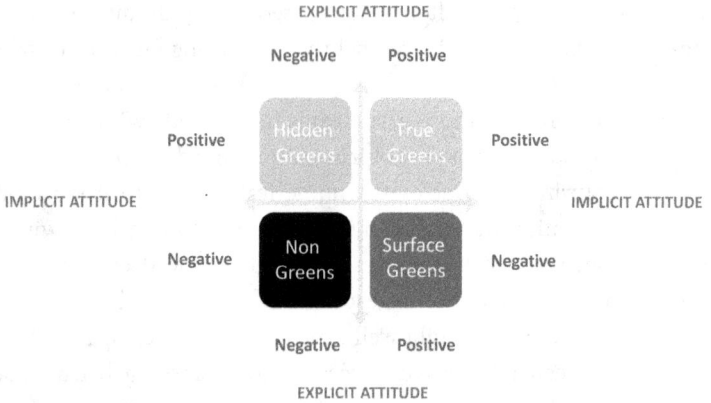

Figure 6.4. Example of implicit attitude/explicit attitude segmentation analysis. (from *Beattie* 2010)

compared to self-report measures. It may, therefore, provide us with a simple diagnostic tool to test the public's actual readiness to go green, in the fight against climate change, and this could turn out to be very important indeed. One could imagine redoing the segmentation analyses of DEFRA and other leading organisations where one attempted to profile the population in terms of both explicit and implicit attitudes rather than relying merely on what people say.

If we really do have a 'divided self' when it comes to our underlying attitudes towards the environment, then this could be critical in the battle against climate change. After all, many of us say that we know that we need to adapt our behaviour as consumers in the light of the threat of climate change, but then actually do nothing. Until we start to promote low carbon products and low carbon lifestyles in a way that impinges on our automatic, unconscious system, little may actually change in this regard. We cannot leave choice of low carbon products solely to reason and reflection, it could be far too late.

As Kahneman (2011) himself has noted, System 2 (the system of reason and reflection) can be very lazy; indeed, it leaves a great deal to System 1, and System 1 is currently prioritising well-known brands and value brands over those with the right environmental properties. System 1, in the domain of consumption, is directing us to choose those things that we have been taught to value – big brands (and status) and economical brands (and money) – rather than environmental brands. This may well need to change. By all means, let's continue to write (and read) the editorials in the quality newspapers about climate change and what we must all do. But at the same time let us think about how to promote low carbon lifestyles as something to do with a new sort of societal status, fun, sexy, necessary, caring, cooperative, clever, perceptive, confident, a must have, the next big thing, a new revolution, in a way that System 1 might notice and build new associative networks.

And, if we have to borrow from the years of (chilling) success of the tobacco industry in promoting smoking (who were brilliant at this sort of thing, see Beattie 2018a), and learn from the way they used associative networks, then so be it. At least, we will then know that it was good for something. They knew that there were two levels of processing (one conscious and one unconscious), two systems, two separate but interacting selves. They were on to something, and they exploited it shamelessly. The rest of us are slowly waking up to this idea.

FURTHER READING

Allport, G. W. (1935). Attitudes. In C. Murchison (Eds.), *Handbook of Social Psychology*. Worcester, MA: Clark University Press.

Beattie, G. and McGuire, L. (2018). *The Psychology of Climate Change*. London: Routledge.

Fazio, R. (1990). Multiple processes by which attitudes guide behavior: The MODE model as an integrative framework. *Advances in Experimental Social Psychology* 23: 75–109.

Greenwald, A. G., McGhee, D. E. and Schwartz, J. L. (1998). Measuring individual differences in implicit cognition: The implicit association test. *Journal of Personality and Social Psychology* 74: 1464–1480.

Kollmuss, A. and Agyeman, J. (2010). Mind the gap: Why do people act environmentally and what are the barriers to pro-environmental behaviour. *Environmental Education Research* 8: 239–260.

7

CAN MODELS OF ANXIETY
HELP US UNDERSTAND
CLIMATE ANXIETY?

Anxiety has been well researched in psychology and there are good models of how it operates. What can we learn from this work when we are trying to understand climate anxiety? Can it help us frame some of the important issues and think about some of the brain mechanisms that are likely to be involved? There are several clinically identifiable types of anxiety disorder. Do they have any characteristics in common with climate anxiety? Is climate *anxiety*, in fact, the right term?

Anxiety has always been with us (climate anxiety is a relative newcomer). We all know what anxiety feels like. I certainly do.

It was often said throughout the latter half of the twentieth century that we were living in the age of anxiety because it was so prevalent. And that was before the public fully appreciated the existential threat of climate change, as well as all the other global crises that we currently face (like global conflicts and the ever-present threat of nuclear war). Perhaps, in earlier times, we were more naïve or had less information available to us about the possibility of disputes and conflicts going nuclear. I, for one, never fully appreciated how close we really came to a nuclear war between the USA and the Soviet Union in October 1962 over the Cuban Missile Crisis. There are few

DOI: 10.4324/9781032631882-9

hiding places from threat these days and consequently more anxiety generated, and the threats are more serious. They are not going to be sorted out with a few presidential telephone calls, a bit of sabre rattling, and a few secret deals.

To gain some insight into climate anxiety, we may want to start with the concept of anxiety itself. Anxiety is one of several anticipatory threat-related emotions (Bohm 2003). We get anxious about something (but quite often about nothing specific) and then (if you're lucky) it slowly fades; the experience is not very pleasant. Indeed, Freud wrote, 'I have no need to introduce anxiety to you. Every one of us has experienced that sensation, or to speak more correctly, that affective state, at one time or another on our own account' (1949: 440). It would seem to be one of the most pervasive and ubiquitous of human emotions across cultures (Sarason and Sarason 1990).

Anxiety is similar to certain other negative emotions, particularly fear. The feeling in both emotions is quite similar, but the two emotions are different in important ways (Zeidner and Matthews 2010). Fear is a biologically adaptive physiological and behavioural response to a clear and present danger with a specific, identifiable stimulus or entity responsible for the emotion (a snake, a man in the shadows, a lightning flash, a snarling dog). This emotion prepares you for fight or flight. This 'fight or flight' response is part of our evolutionary history. Anxiety is also thought by many psychologists to be biologically adaptive, it orients the individual to the anticipation of perceived danger or threat, acting like an 'early-warning radar', putting you on guard. But it can be dysfunctional; it responds to the perception of threat and this perception may be faulty.

Anxiety is more 'diffuse' than fear – it is sometimes triggered by a specific upcoming event (an exam, meeting your boyfriend's parents, going to the doctor's), but sometimes the cause of the anxiety is harder to pinpoint. You wake up feeling anxious and you start thinking about what could possibly be behind this unpleasant feeling, and occasionally you 'discover' what's bothering you and making you anxious. But sometimes the source of the anxiety can be quite elusive. I certainly have said to my partner on a number of occasions, 'I've woken up feeling anxious but I've no idea why'. She has a standard reply. 'You're meant to be the psychologist, work it out.'

The last time this happened my partner had just found a very large lump in her breast, which she assured me would be fine. She was referred to a cancer clinic. She was genuinely very relaxed about the whole thing. I didn't feel anxious immediately, I was reassured by what she told me. But that night, I dreamt that we were on a walk through some woods. It was autumn somewhere in England (the leaves were turning brown, fluttering down, crispy underfoot). She was slightly in front of me. Then suddenly a large black puma (somewhat incongruent in this English autumnal setting) sprang out in front of us with its teeth bared. It started to move towards us, slowly at first, stealthily, but then began speeding up. It was about to pounce on her. I ran SCREAMING towards the puma to frighten it away, blocking it from my partner, protecting her. I was shouting 'WAAAAA' in my sleep, very loudly. I know what I was shouting because it was so loud that it woke both of us up. I went back to sleep, but when I finally woke up properly, I was decidedly anxious. I wasn't sure whether it was the possibly cancerous lump (which turned out to be a benign cyst), or my very vivid experience with the puma which had left me a little shaky (it was very vivid and realistic), or the fact that sleep can sometimes not offer any refuge from the cares of the world. These three things are all related, of course, in this context, but separate.

You don't have to be a Freudian psychologist to appreciate that the black puma was symbolically representing the cancer (dark, hidden, stealthy, and then the sudden attack on my partner with great speed, the lump was already very large), but if the cancer hadn't been represented in this way in my dream would I have been quite so anxious when I woke up? My unconscious was representing the original source of my anxiety in a striking and fearful way. Was that particular representation rather than all the other possible representations making my anxiety state worse? Or was it the vividness of the dream that had alarmed me? I never shout out in my sleep (I am told). In other words, was the cause of the anxiety the original stimulus (the lump) or the various psychic consequences? I wasn't sure. Or was it all three in combination?

'Did you come up with anything?' she asked me later.

'It's a work in progress', I replied. I didn't want her to think that I had confused the symptoms with the cause. But I kept thinking that these symptoms of anxiety realised in the dream might have exacerbated the feeling of anxiety.

That's my working hypothesis, kept secret until now.

Anxiety, like fear, is experienced through cognitive symptoms, like worry and intrusive thoughts, but also through *somatic* symptoms, like a fast-beating heart, perspiration, and butterflies in the stomach, all classic signs of the so-called 'fight or flight' response. Zeidner and Matthews (2010) say that the subjective feeling of anxiety is often more difficult to tolerate than fear because it is 'more pervasive and draining'. With fear, they say, you might know what to do, i.e. whether it is better to fight or to flee in the circumstances; with anxiety, the person is often uncertain how to act. The two emotions also have different neuroanatomical substrates, with fear centred on the amygdala and anxiety in the bed nucleus of the stria terminalis.

A major issue in anxiety research is the distinction between *trait anxiety* and *state anxiety*. Spielberger (1966) in some classic early research differentiated temporary feelings of anxiety (state anxiety) from a person's more general tendencies towards anxiety (trait anxiety). Those individuals who are more easily made anxious than others are described as having high trait anxiety. Given a certain level of threat, those with high levels of trait anxiety will experience higher levels of state anxiety than those with lower levels of trait anxiety. In other words, the level of state anxiety experienced in any situation will be a function of the perceived threat *and* underlying personality (particularly level of *neuroticism* which reflects degree of emotional stability). There is therefore the danger of dismissing anxious individuals as just neurotic (troubling given the connotations the concept has acquired in everyday use). This may also be a significant issue with those who suffer from climate anxiety.

Worry and *concern* are also threat-related states where you see the future as uncertain and potentially dangerous, but they are generally considered to be more thought-based and less intense than anxiety, in other words more cognitive and less genuinely emotional (Borkovec et al. 1983; van der Linden 2017). In some more general models of anxiety, they are seen as components of anxiety, part of the *secondary elaboration* to deal with the perceived threat (Beck and Clark 1997). A degree of anxiety (as well as worry and concern) can be adaptive in certain circumstances, these authors argue, guiding you to respond appropriately to a threat that you recognise. You get anxious about an upcoming exam, so you do more revision, and you pass successfully, but whether it helps or hinders depends upon the level of anxiety suffered. The level of anxiety is critical to the outcome. If you get too anxious about the exam, you might just freeze.

Anxiety can be abnormal as well as normal. Zeidner and Matthews (2010) explain that clinical anxiety is more intense than normal anxiety and it involves distortions in thinking where people may misinterpret events and perceive a threat where none actually exists. Those suffering from clinical anxiety might be highly biased in their perception of threat. Anxiety can be extremely harmful and overwhelming, involving conditions like generalised anxiety disorder (GAD), phobias, and post-traumatic stress disorder (PTSD), all clinically recognised anxiety disorders included in DSM-5. Generalised anxiety disorder refers to longer-lasting, non-specific and 'free-floating' anxiety, which is not tied to any specific threat or event, and which causes significant problems for functioning in daily life. The symptoms are feelings of stress or worry that affect your daily life and are difficult to control, difficulty sleeping, getting tired easily, feeling tense, stomach problems, heart palpitations, feeing lightheaded or dizzy, and low mood or depression. It may also be associated with sudden attacks of anxiety and panic (panic disorder). Phobias are an overwhelming fear of something specific, often in response to specific stimuli (snakes or spiders) or particular physical contexts (open or confined spaces). These generate intense fear responses (with sweating, trembling, shortness of breath, rapid heartbeat, pain or tightness in the chest, butterflies in the stomach, headaches, feeling faint, dry mouth, confusion or disorientation, fear of losing control, feelings of dread, etc.). PTSD is an enduring anxiety caused by a traumatic and horrific event characterised by re-experiencing of the trauma through flashbacks (vivid intense memories that do not fade with time), nightmares, repetitive and distressing images or sensations, physical sensations such as sweating or trembling, emotional numbing, hyperarousal and hypervigilance (feeling 'on edge' and being on guard all the time), irritability, angry outbursts, insomnia and difficulty concentrating.

Some of the most influential research on anxiety over the past thirty years (influential in terms of both amount of research and implications for treatment) has considered anxiety disorders from what is called an information processing perspective. This derives from cognitive psychology and follows the so-called 'cognitive revolution' in psychology with the emphasis on the mind as an information processor like a type of computer (with the proviso we discussed earlier). Underpinning this work, is the understanding that 'the type of emotional information and the manner in which it is processed are crucial factors in the etiology, maintenance and treatment of anxiety disorders' (Beck and Clark 1997: 49). According to Aaron Beck, one of the world's leading clinical psychologists, a selective bias for threat information in attention, interpretation and memory is a central characteristic of all anxiety disorders. This combines with two other features. The first is the individual underestimating their 'personal coping resources'; the second is them underestimating 'the safety or rescue features in the environment'. From Beck's point of view, anxiety is the result of certain identifiable *cognitive biases* with too much attention to potential threat, and too little attention to one's own coping resources and the benign features of our environment that could possibly assist. According to Beck, those who suffer from anxiety have a different sort of unconscious cognitive bias towards threat information (as a result of their past experiences).

Beck explicitly acknowledges the role of unconscious and automatic (System 1) processes in the very first stages in the perception of threat, including the orienting mode:

> The function of this 'early warning detection system' [the orienting mode] is to identify stimuli and assign an initial processing priority through the allocation of attentional resources. Moreover, the orienting mode is most likely to assign information processing priority to stimuli or situations that threaten the survival of the organism.
>
> (Beck and Clark 1997: 49)

However, processing of the threat at this stage is relatively undifferentiated according to Beck – focusing on just affect (and categorised as positive, neutral or negative) and personal relevance. Following this initial stage, the orienting mode leads to the *immediate preparation* and activation of the *primal mode* to maximise the chances of survival. Activation of the threat *primal mode* results in a coordinated goal-directed strategy aimed at minimising danger and maximising safety with a number of primal responses:

1. Autonomic arousal – the preparation for fight or flight.
2. Behavioural mobilisation   to prepare for escape and avoidance behaviour.
3. Primal thinking – a narrowing of cognitive processing onto the threat as well as the production of repetitive, involuntary, automatic thoughts and images involving threat and danger.
4. Feeling of fear which motivates the individual for action.
5. Hypervigilance for threat cues.

Beck writes: 'In sum once the primal threat mode is activated, it tends to dominate the information processing apparatus, thereby blocking off other secondary, more constructive or reflexive modes of thinking' (Beck and Clark 1997: 52). In the terminology we have used in previous chapters, because System 1 is fully engaged with the perceived threat, the *logical* System 2 is being blocked from making its full potential contribution to understanding the nature of the threat, at this stage at least. The primal threat mode does some partial appraisal based on incomplete information, but because the cognitive processing is constricted and narrow, this leads to biases and inaccuracies:

> The anxious individual engages in selective abstraction, becoming hypersensitive to the potentially harmful aspects of a situation but ignoring its more positive features ... Finally, there is an overestimation of the probability and severity of the threatening situation resulting in the catastrophic thinking that so characterises anxiety disorders such as panic.
>
> (Beck and Clark 1997: 53)

A second product of primal mode activation is the occurrence of *negative automatic thoughts* involving themes of threat and danger. This primal mode leads to the final stage in the cognitive model of anxiety – that of *secondary elaboration*. This is the slow, effortful and conscious System 2, with a more reflective consideration of what is occurring and one's coping resources. Is the threat as serious as it first seemed? Or not? This is where *worry* comes in. This is one of the core *mechanisms* that are used in this stage of secondary elaboration in this most influential theory. According to Borkovec and colleagues (1983), worry is:

> A chain of thoughts and images, negatively affect-laden and relatively uncontrollable. The worry process represents an attempt to engage in mental problem-solving on an issue whose outcome is uncertain but contains the possibility of one or more negative outcomes. Consequently, worry relates closely to fear processes.
>
> (Borkovec et al. 1983: 10)

This cognitive model of anxiety has clear implications for treatment, which is why it has been so influential both in terms of theory and in practice. Beck writes:

> The treatment of anxiety must involve the deactivation of the automatic, hypervalent primal threat mode and a strengthening of the elaborative, strategic processes involving the activation of more constructive, reflective modes of thinking ... Furthermore because the model proposes that the threat-meaning assignment is the *sine qua non* of anxiety disorders, effective treatment must modify the threat appraisal process.
>
> (Beck and Clark 1997: 55)

So how does this work in practical terms? You treat anxious patients' 'coping strategies', for example, by countering the primal threat mode with 'more elaborative, strategic processing of information resulting from the activation of the constructive, reflexive modes of thinking'. In other words, you get them to bring reflexive logical thinking into their appraisal of the threat and of their coping resources. This is all done consciously, and worry is an important aspect of this process. The goal is that you want anxious patients to reason sufficiently to understand that the threat is not as serious as it first seemed, and to re-evaluate the availability and effectiveness of their own coping resources, hopefully to conclude that they have the resources to deal effectively with it.

So how does any of this apply to 'climate anxiety'? Climate anxiety is clearly serious and although it's not a separate psychological disorder in terms of inclusion in DSM-5, it can become highly clinically relevant if it is intense enough to start interfering with the ability to sleep, work or socialise (Clayton 2020). The fact that it is not included in DSM-5 has many serious disadvantages (as I have mentioned a number of times), in that people may define or operationalise climate anxiety differently in different studies (and we will soon see why this is an important issue). They may be researching and writing about different psychological states. It may also be harder to access appropriate mental health support without formal clinical recognition (van Valkengoed 2023).

The prevalence of climate anxiety varies depending upon the specificity of the definitions used. Some researchers take high levels of worry as constituting climate anxiety, others use the specific term 'anxiety' in their research. Gregersen (2024) reported that 56% of young adults in Norway report being *worried* or *extremely worried* about climate change. Ogunbode et al. (2022), on the other hand, reported that 28% of Norwegian university students report feeling *very* or *extremely anxious* about it. So, does climate anxiety affect a quarter or half of young adults? If it has to include clinically significant functional impairment, the figures are lower still. Whitmarsh et al. (2022) reported that 9% of young people in the UK under 30 years of age suffered from climate anxiety that impacted on sleep or concentration. So where do we draw the line for climate anxiety? High levels of worry? High levels of anxiety? High levels of anxiety that impact on daily life? The gap between 9% and 56% is a very big one, so we need to be clear about what exactly we mean when we say that someone is suffering from climate anxiety.

But there is another much more pressing conceptual issue here about climate anxiety when you consider the writing of the extremely influential Aaron Beck. It is clear how Beck thinks about anxiety, and it is worth repeating.

> The anxious individual engages in selective abstraction, becoming hypersensitive to the potentially harmful aspects of a situation but ignoring its more positive features … Finally, there is an overestimation of the probability and severity of the threatening situation resulting in the catastrophic thinking that so characterises anxiety disorders such as panic.

But how can you be *hypersensitive* to an existential threat like climate change? How can you '*overestimate* the probability and severity of the threatening situation resulting in *catastrophic thinking*'. Just read the science (the latest IPCC report will do), and then explain to me how we can use words like 'hypersensitivity', 'overestimation of threat' and 'biased catastrophic thinking' when we are talking about climate change. It is almost as if the classic models of anxiety were developed in safer times (even 'the age of anxiety'!) and were appropriate for this different age, whereas the existential crises we face now are quite literally just outside our door. The late twentieth century may have been described as the age of anxiety, and the cognitive behavioural therapy developed by Aaron Beck may have been suitable for that age, but should we be thinking along the same lines now and basing our understanding on these kinds of propositions? How can we 'overestimate a threat' like climate change? Isn't 'catastrophic thinking' entirely appropriate here? And if it is, do we need to rethink the very nature of anxiety for our modern and troublesome world?

But we also mustn't throw the baby out with the bathwater. Beck explains in his model that anxiety affects both thinking and the body, cognitions and emotions, and that worry is part of the mechanism of anxiety, but it's only a part, it's not the whole thing. Anxiety affects both systems of the brain, System 1 and System 2, the unconscious and automatic part and the slower rational part. Beck's model has helped thousands deal better with anxiety. We need to bear in mind how climate anxiety might differ from 'normal' anxiety but not ditch the whole thing. After all, the brain and its functioning are the constants here and we are seeing evidence of the importance of System 1 for attitudes, behaviours and now anxiety.

FURTHER READING

Beck, A. T. and Clark, D. A. (1997). An information processing model of anxiety: Automatic and strategic processes. *Behaviour Research and Therapy* 35: 49–58.

Clayton, S. (2020). Climate anxiety: Psychological responses to climate change. *Journal of Anxiety Disorders* 74: 102263.

Gregersen, T. et al. (2024). How the public understands and reacts to the term 'climate anxiety'. *Journal of Environmental Psychology* 96: 102340.

Whitmarsh, L. et al. (2022). Climate anxiety: What predicts it and how is it related to climate action? *Journal of Environmental Psychology* 83: 101866.

Zeidner, M. and Matthews, G. (2010). *Anxiety 101*. London: Springer.

Part 3

CONTESTING CLIMATE ANXIETY/
CONTESTING CLIMATE CHANGE

8

WHY DO PEOPLE CLAIM CLIMATE CHANGE IS FAKE NEWS?

Climate change deniers are more than a minor irritant. They create uncertainty about climate change when there should be none. They say, 'How can you have climate anxiety about something that isn't happening?' They have a platform, and they persuade others of their position, at odds with science, at odds with the truth. Climate change mitigation requires a concerted and unified *global* effort. It must involve us all. Climate deniers help prevent this from ever occurring. They, therefore, impact greatly on the experience and prevalence of climate anxiety. There are some especially significant climate deniers like Donald Trump who, in my opinion, require special psychological analysis. Why does he say these things?

DRILL BABY DRILL!

There are many well-known climate change deniers, but none is more famous or influential than President Donald Trump. In 2024, as I was working on this book, Trump was in the run-up to the 2024 presidential election, and he was still calling climate change 'fake news' and 'a great scam'. In October 2024, he was still neck and neck with the Democratic candidate Kamala Harris. They were saying that the election could go either way. Many in the United States and

DOI: 10.4324/9781032631882-11

elsewhere were extremely worried at the prospect of Trump being re-elected as president. His views on climate change were one major cause for concern but there were many others (including his apparent wavering support for Ukraine in its war with Russia and his cosy relationship with Vladimir Putin).

I returned to this particular chapter on 6 November. Trump had just won; Kamala Harris hadn't yet conceded defeat, but she was about to. The image of Trump shouting to his supporters 'we're going to drill, baby, drill' and 'frack, baby, frack', accompanied by some pelvic thrusting (I'm sure that's what I saw!), has now been implanted in my worrisome mind.

Trump's views are extremely influential. Researchers using artificial intelligence (AI) and network analysis analysed 7.4 million tweets posted by roughly 1.3 million people on the social media platform X (formerly Twitter) in the USA between 2017 and 2019 (Gounaridis and Newell 2024). These social media posts were coded as 'for' (belief) or 'against' (denial) climate change using AI. Over half of the tweets denied that climate change was real, and climate change denial was highest in the central and southern states of the USA, particularly among Republican voters. The researchers found that political affiliation was the best predictor of climate change denial, and that Donald Trump was the strongest influencer in this network. He doesn't believe that there is a climate crisis. To him, climate anxiety would be a sure sign of neurosis. How can you be anxious about something that's not real?

Trump has been consistent in his climate change denial, going back to 2017 and his first election as president. That was a critical year for climate change. On 20 January, Trump was elected the 45th President of the United States, and later that year, the Fourth National Climate Assessment Report was published by the US Global Change Research Program. Two monumental events for the 'debate' on climate change. Trump said that he would cancel the Paris Climate Agreement within 100 days of taking office; he signed an executive order in March 2017 that reversed the Clean Power Plan that required states to regulate power plants. He described anthropogenic climate change as 'fake news' and 'fictional'.

The Fourth National Climate Assessment Report was yet another report that bolstered the scientific consensus on climate change, but this one was 'the authoritative assessment of the science of climate change', with a focus on the United States. The fact that the focus was the US was very important. One major psychological issue with climate change is that it is often perceived to be primarily about other places and other times, rather than impacting on our own lives. The belief is that it will affect more distant locations (called 'spatial bias') and not our own, and future generations rather than ours ('temporal bias'). Indeed, in research carried out in the UK in 2017 with some colleagues (and detailed in the next chapter), I found clear evidence for these biases – people reckoned they had a 48.1% probability of being personally affected by climate change, but other people living today had a 65.3% probability of being affected. The respondents also reckoned that 82.8% of future generations would be affected (Beattie et al. 2017).

Large sections of the population of the US seemed to assume that they would be immune to the whims of climate change (if it exists at all), and Donald Trump in his first election campaign tapped into these beliefs and reinforced them. He became the cheerleader for climate change denial. A significant proportion of the American population seemed to believe that it didn't really concern them (except perhaps in terms of what they might have to pay, in the light of the Paris Climate Agreement). Many, including the new president himself, described it as a 'scam', and this message played very well in his campaign in those states which had been decimated by the decline of the coal industry. He tweeted on 1 November 2012, 'Let's continue to destroy the competitiveness of our factories & manufacturing so we can fight mythical global warming. China is so happy!', and on 15 February 2015 he tweeted, 'Record low temperatures and massive amounts of snow. Where the hell is GLOBAL WARMING?' 'Where the hell is global warming when you need it?' became a recurrent slogan.

'Right here, right now', was the answer from the Fourth National Climate Assessment Report. This report read:

> Global annually averaged surface air temperature has increased by about 1.8° F (1.0° C) over the last 115 years (1901–2016). This period is now the warmest in the history of modern civilization ... it is extremely likely that human activities, especially emissions of greenhouse gasses, are the dominant cause of the observed warming since the mid-20th century ... in addition to warming, many other aspects of global climate are changing, primarily in response to human activities. Thousands of studies conducted by researchers from around the world have documented changes in surface, atmospheric, and oceanic temperatures; melting glaciers; diminishing snow cover; shrinking sea ice; rising sea levels; ocean acidification; and increasing atmospheric water vapor.
>
> (Fourth National Climate Assessment 2018: 10)

The upshot of these changes for the United States was well documented in the report – it describes an increase in extreme weather events with heavy rainfall, a higher frequency of heatwaves, larger forest fires in the western United States and reduced snowpack in Alaska affecting water resources in the western United States. The report warned that, 'assuming no change to current water resources management, chronic, long-duration hydrological drought is increasingly possible before the end of this century'.

This was a balanced and authoritative scientific assessment, but science, of course, works on the principles of scientific testing and prediction and probability. Very few things in life are actually certain. So, the report said that 'it is *extremely likely* that human activities, especially emissions of greenhouse gasses, are the dominant cause of the observed warming since the mid-20th century'. Science is based on probabilities and the report, therefore, goes to the trouble of explaining these key probabilistic terms with a glossary. They explained that 'likelihood' is the 'chance of occurrence of an effect or impact based on measures of uncertainty expressed probabilistically'. They also explained that 'extremely likely' means that it has a 95%–100% chance of occurring. Scientists understand the full significance of this. But, of course, critics, cynics, extreme optimists, those with a vested interest, the President of the US, seized on these probabilistic terms. 'It's not certain', they would say. Why should we change our behaviour, our values, our culture, our economic position in the world for something that is just *likely*? Okay, extremely likely. This doesn't mean that it's going to happen *for sure*. If we change our coal and oil industries, I'll tell you what, we're going to surrender our economic position to China, and that is *for sure*.

The year 2017 was a year of non-science, and 'fake news' and discussions that weren't.

Towards the close of that year, things started hotting up. Trump was on vacation in Mar-a-Lago in West Palm Beach. The sun was shining but then there was a cold snap on the north-east coast of the United States over Christmas and New Year. Dogs froze to death in their kennels. Trump was obviously delighted. 'In the East, it could be the COLDEST New Year's Eve on record. Perhaps we could use a little bit of that good old Global Warming that our Country, but not other countries, was going to pay TRILLIONS OF DOLLARS to protect against. Bundle up!' he tweeted gleefully on 28 December. And then in the New Year, Florida had its first snowfall in nearly three decades. Frozen iguanas were dropping from the trees. House-owners in the Sunshine State were warned to leave them alone until they defrosted. The most powerful man in the world had evidence that climate change was a total hoax – frozen iguanas that you have to defrost in Florida. 'Global warming is a SCAM', he roared. But Trump didn't seem to understand the difference between 'climate' (the bigger picture across time) and 'weather' (the smaller, more localised picture, with a whole series of fluctuations and change). On 1 November 2011, he had tweeted, 'It snowed over 4 inches this past weekend in New York City. It is still October. So much for Global Warming.'

But Donald Trump is always a stark and ubiquitous reminder that we all don't think alike when it comes to climate change. There is clearly a great divide between believers and non-believers, between Republicans and Democrats, between the right-wing and left-wing press. The statistics on this divide fuelled by ideological position are striking. Hoffman writes that,

> In 1997, 47% of Republicans and 46% of Democrats thought that climate change was already happening, in other words virtually identical percentages. By 2008, the figures had diverged dramatically with fewer Republicans holding this view (down to 41%) but with far more Democrats than previously expressing this position (up to 76%). By 2013, the respective figures were further apart still, 50% and 88% respectively.
>
> (Hoffman 2015)

Hoffman says that the cause of this polarisation was the Kyoto Protocol, which was the first international agreement to reduce greenhouse gas emissions. Media attention on the economic implications of climate change rose dramatically in the years following. McCright and Dunlap (2011) reported that there were 166 documents critical of the science of climate change in 1997 alone; 107 climate change denial books were published between 1989 and 2010. Most of these, according to Hoffman, were linked to conservative think tanks, and somewhat tellingly, 90% did not go through a peer review process (the very bedrock of science itself).

The Kyoto Protocol had major implications for the energy sector and industry in the US, so a counter-campaign was mounted. The same thing happened when the conclusive science linking smoking and lung cancer emerged. Both were turned into 'scientific debates', which allowed both viewpoints to flourish (Beattie 2018a). For decades, it seemed there were two tenable positions on climate change supported by climate science. On one side, you had the climate change believers, on the other, climate change deniers. You can therefore choose which side to take – that was the implicit message. The BBC and other broadcasters had a 'duty' maintained over many years to have both perspectives represented in any 'balanced' discussion on climate change. This media balance reinforced the notion that there was serious doubt about the science of climate change and many people liked this message (no existential threat, no change necessary, continue as before) and climate change sceptics seemed to be everywhere.

Greta Thunberg was trying to remove all doubt with her simple message, with no ambiguity and no window dressing. But the problem is that too much fear in any message ('Our house is on fire') is also not an effective way of gaining compliance – if the recipient of the message doubts whether there is anything they can do to resolve the threat and reduce the fear. You need to know that you have the power to do something (self-efficacy) and that your response will make a difference to mitigate the effects (response efficacy). If you have doubt about either or both these things, you find other ways of dealing with the fear.

Greta Thunberg now crystallised (and personified) this extreme fear message about climate change, but ordinary citizens often felt powerless about whether they could do anything which would actually make a difference. They developed an aversion to Greta and her message. In situations like this they may well attempt to avoid the message altogether – refusing to read articles about climate change or watch news items about it (or even turning off the TV when she appears on screen), or more subtly not attending to certain parts of the article even when they're right in front of you on a computer screen (Beattie et al. 2017). These 'certain' parts to be avoided being the remarkable consensus on the science and the scary consequences. If the climate change sceptic arguments are presented as well, i.e. those arguments that raise severe doubts about climate change, then some people find solace in these. They attend more to these sections – their eyes automatically and seemingly unconsciously drawn to them. That is the information they then process to build their representation of the situation, to confirm their long-held views – namely that climate change won't affect *them* – just other people, overseas or in the future (or preferably both). You may even conclude that these climate change reports are attempts by *others* to manipulate you.

The 'debate' is still ongoing. In October 2024, presidential candidate Trump was still describing the climate crisis as 'one of the great scams'. He had told an audience in Wisconsin that under the 'green new scam', Democrats 'wanted to rip down all the buildings in Manhattan and they wanted to rebuild them without windows' (*The Guardian*, 1 October 2024). *The Guardian* quotes Brett Hartl, political director at the environmental non-profit Center for Biological Diversity Action Fund: 'It's obscene that communities … are suffering and dying from the reality of the climate emergency while Donald Trump denies that it even exists.' Trump made these pronouncements whilst on his way to several fundraising events in oil-rich Texas, thrown by oil industry executives.

But Trump and other climate sceptics don't just deny the evidence, they go beyond this – they make it a complete counter-narrative – they dispute not just the science but say that they can expose the architects behind this shady science, this 'fake news' – the IPCC ('mere puppets'), certain governments, the Chinese – all out to ruin Western economies, and particularly out to destroy American business. It's not hard to understand the alignment between climate change denial and particular political parties.

A few years ago, I remember watching an extraordinary sight on my television. There was a young girl in pigtails in a casual grey checked top holding some notes in front of her. She looked younger than her 16 years. The screen behind her read 'World Economic Forum'. It was the incongruity between the caption and this image of a child who hadn't bothered to dress up for the cameras that grabbed your attention. There was silence, she adjusted the microphone and then she started to speak. She sounded confident; she didn't smile. 'Our house is on fire … I'm here to say our house is on fire …' She paused after each short sentence, the sentence was repeated, her eyes flicked to the left and right as if monitoring for feedback, but you felt that any feedback wouldn't influence either the message or the delivery. It was a wary sort of watching. Then there was another pause.

> According to the IPCC we are less than twelve years away from not being able to undo our mistakes. In that time unprecedented change in all aspects of society needs to have taken place, including a reduction in our CO2 emissions of at least fifty percent.

Her message was that clear and that uncomfortable, but it aligns perfectly with the science. If you accept her message, if you believe her, then is it little wonder that people can get anxious about climate change. But there is no doubt in my mind that some portion of the anxiety is attributable to the fact that we know that many other people aren't convinced about the role of human activity in climate change (or even that it's changing) because of these alternative 'facts' and alternative discourses presented to them. The problem is that we need the level of consensus that is found between climate scientists replicated in the public as a whole if we are to be effective in our efforts to mitigate the effects of climate change.

Van der Linden (2015) demonstrated that participants exposed to conspiracy videos about climate change ('it's all fake') were less likely to think that there was widespread scientific agreement on the role of human activity in causing climate change and less likely to engage in activities to help other people (like donating to a charity or volunteering for charitable activities). In other words, these alternative theories pull us inwards and make us less interested in helping our fellow citizen.

WHAT COMES NATURALLY

Of course, there is something else that may be relevant here when we consider Trump's views on climate change. He is a proven liar. Many politicians do lie, of course, but Trump is in a class of his own (Beattie 2024a). According to the *Washington Post*, he made 30,573 false or misleading claims in his four years as president. They increased year on year, from six per day in his first year to thirty-nine in his final year (Beattie 2024b).

Although most presidents have lied to the public at some time or other, none have lied with this sort of frequency or told lies that are so wide ranging. Trump lies about everything, even about trivial things, but many are self-aggrandising lies ('nobody builds better walls than me', in reference to a new wall planned for the Mexican border, but when has he ever built a wall before good or otherwise?), or more egregious lies (like the lie about the outcome of the 2020 presidential election), demonstrably contrary to the facts, with serious consequences for public trust. And these lies can cut through. Research by Arceneaux and Rory Truex in 2023 analysed how belief in the 'big lie' about the outcome of the 2020 presidential election, when Trump claimed victory and refused to accept defeat, varied over the next forty days after the election. The lie was pervasive and effective. Half of ordinary Republican voters believed that the election was stolen, and this belief persisted over that whole period. The lie boosted Republican supporters' self-esteem.

Politicians who lie have an enormous advantage over truth-tellers. If you can successfully embellish the truth or construct a new reality, that can always be more interesting and engaging than the truth (Danesi 2020). The truth may be a bit dull and uninspiring; the lie can be whatever you want it to be. You know what your audience wants to hear. And besides, what is there to lose? Politicians know that lying is part of our everyday lives (Beattie 2016). We all do it! Research in psychology using lie diaries tells us that people lie on average twice a day (DePaulo et al. 1996). Many are harmless 'white' lies told for the benefit of others, but some are not so harmless and told for the benefit of the liar themselves. Some people get significant pleasure from telling such self-centred lies. Psychologists call this 'duping delight' (Ekman 2001). It confuses the recipient of the lie. They expect to detect signs of guilt or anxiety, instead all they see is a faint smile of satisfaction. The liar gets away with it (Beattie 2024a). That smile could mean anything.

We start lying early in life – between 2 and 3 years of age. Charles Darwin in 1877 caught his son William Erasmus lying at that age. He had eaten some forbidden pickle juice and lied about it. Darwin was surprised by his son's evident pleasure in lying (Darwin 1877). Following Darwin, developmental psychologists studied how children learn to tell lies, and were particularly interested in the moral development of the child (see Hartshorne and May 1929). However, lies and lying are also complex cognitive skills and two cognitive factors have been identified which are thought to play a major role in the ability to lie effectively (Sai et al. 2021). These are Theory of Mind and Executive Function.

Theory of Mind refers to the ability to understand and attribute mental states, such as beliefs, desires, intentions, emotions and knowledge, to oneself and others and, critically, to recognise that other people have thoughts and feelings that might be different from one's own. This cognitive skill is crucial for social interaction and communication, as it enables individuals to predict and interpret the behaviour of others based on their mental state. Having a theory of mind allows us to empathise with others, understand their intentions, and make more accurate predictions about their behaviour. It's also critical to lying. In order to tell a lie, children must represent and differentiate between the mental states of themselves and the recipient of the lie, and then use words (or actions) to instil false beliefs in this other person using the lie (Talwar and Lee 2008). In order to tell a successful follow-up lie and maintain consistency with the initial lie, children must also be able to infer what false beliefs they have implanted in the mind of the recipient of the lie (Talwar and Lee 2008). Executive Function, on the other hand, refers to a set of higher-order mental processes and cognitive abilities that are crucial for goal-directed behaviour, decision-making, planning and self-regulation. It manages and coordinates different cognitive functions to achieve desired outcomes effectively.

A meta-analysis of forty-seven papers on the development of lying published in 2021 found that both Theory of Mind and Executive Function are positively correlated with frequency of lying in children (Sai et al. 2021). This association between Executive Function and lying was also strongest for self-protective lies (lies to conceal transgressions to avoid negative consequences) and self-benefiting lies (lies to get a strategic advantage) rather than for 'white' lies (prosocial lies in order to be polite – 'you look great in that dress' etc.) or lies to protect somebody else.

In other words, the brightest children in terms of their Theory of Mind ability and Executive Function lie most frequently and most selfishly. Those smart children, with a good working memory, flexible in their thinking and able to stop themselves automatically telling the truth through good inhibitory control, lie effectively (Williams et al. 1999). But most of us still feel guilty about lying, and that's what stops us in the end regardless of how smart we are. We feel ashamed and embarrassed, remorseful, guilt-ridden. However, individuals with certain types of personality aren't prevented by this. They are drawn to lying, and particularly to self-serving lies, they don't feel remorse like the rest of us. They don't display any emotional nonverbal leakage when they're lying, so they can get away with it time and time again (Cohen et al. 2010; Ekman 2001). Their secret is that they have low levels of empathy, they don't care about other people, they think that they are what really matters (Beattie 2024a). We call these people narcissists; those high in non-clinical psychopathy show similar traits.

SYMPATHY FOR THE LITTLE PEOPLE

Psychopathy is a dimension identified using Robert Hare's (1991) Psychopathy Checklist. Those at the top end of the dimension lie easily, they do not suffer from shame, guilt, or fear of being caught, and importantly they have low levels of affect generally (things that get the rest of us excited have less of an effect on them; things that scare the rest of us scare them less). Psychopaths, including non-clinical psychopaths, are prone to boredom and lie to give themselves a little bit more excitement; they need that buzz from everyday interaction. A few lies can spice things up. They can be charming, and they use lies to be even more charming – to present themselves in particular ways without the truth getting in the way. They have a grandiose sense of self-worth and lies are used to maintain this. They are conning and manipulative and they lie to con and manipulate people out of money or into relationships. They are callous and don't care about the feelings of the other person, the person that they're lying to. They just don't feel the other person's suffering in the normal way. They have low levels of empathy; this makes lying very easy. Some have good jobs; some have great jobs, they're not all isolated loners.

Empathy is a term that is bandied about a lot these days and we all think that we know what it means. But Blair (2005) has argued that it's not a unitary concept but 'a loose collection of partially dissociable neurocognitive systems'. The three main divisions are *cognitive empathy* or Theory of Mind (which we have already come across – the ability to represent the mental states of others to explain and predict their behaviour), *motor empathy* (the tendency to automatically mimic and synchronise facial expressions, postures, and movements with those of another person), and *emotional empathy* (the ability to recognise and respond appropriately to the display of emotions in others). Psychopaths have a problem with *emotional empathy*, and not Theory of Mind (Richell et al. 2003). Children with psychopathic tendencies have an impairment in the recognition of sad expressions (Stevens et al. 2001) and fearful expressions (Blair et al. 2005); adults with psychopathy and children with psychopathic tendencies show reduced autonomic responses to the sad expressions of others (Blair 1999), but they show no impairment in the processing of happy expressions (Stevens et al. 2001).

In other words, the problems in emotional empathy shown by individuals with psychopathy appear relatively selective and seem to involve some dysfunction in the amygdala, that part of the brain central for the processing of these sorts of emotions. The behavioural implications of this dysfunction should be clear. If you cannot recognise when someone is sad or fearful, and have a reduced autonomic response to these emotions, then there are few brakes on your own behaviour. Others might stop when they see that someone is frightened or sad, the psychopath does not. The effects of lying to victims and seeing their distress, similarly, may have little effect; if you can't interpret their negative emotions, or feel their pain, you can lie to your heart's content.

But psychopathy tends to be associated with other personality characteristics, namely narcissism and Machiavellianism (Furnham et al. 2013), and psychologists call these three personality traits the 'Dark Triad'. Narcissism and Machiavellianism are also associated with lying. Narcissists are low on empathy and crave positive attention – they will lie when they must in order to get attention. Narcissists are thought to suffer from extreme selfishness, with a grandiose view of their own talents. In other words, people who are high on narcissism think that they are better than others in terms of many dimensions, including their looks, their intelligence, their creativity and what they do, but according to Twenge and Campbell (2009) they are not. Narcissists also lack emotionally warm, caring, and loving relationships with other people. Narcissists have a need to boast about any achievements or accomplishments in their lives to seek affirmation. They will focus on their physical appearance (amongst other things) and value material goods that can display and communicate instantly their social status relative to everyone else. In social interaction, they will try to make sure that conversations centre on them; in relationships, narcissists will seek out trophy partners that make them look good. Machiavellians manipulate social situations and will use lies to do this; and psychopaths are callous and lack empathy, as we have just considered, and they will be oblivious to the consequences of their lies (Meere and Egan 2017).

Research tells us that self-reported Dark Triad ratings are correlated with several antisocial behaviours, including higher levels of delinquency and aggression in children, as well as risky and sensation-seeking activities (Crysel et al. 2013). The traits which are characteristic of the Dark Triad are thought to be associated with a 'compromised' or 'dysfunctional morality' (Campbell et al. 2009), in that they 'value self' over other 'in a way that violates implicit communal sentiments in people' (Jonason and Webster 2012).

It seems that callousness and lack of emotional empathy are the key attributes underpinning the Dark Triad; entitlement underpins narcissism which is another key attribute. All of these connect to the frequent and expert use of lies. Lies to make you look good, lies to manipulate, and lies without consequence where you don't react to emotions like fear and sadness in the normal way. You don't feel the suffering of any victims.

THE IMPUDENCE TO TELL 'BIG LIES'

Politics was once thought of as a noble art; it was Machiavelli in 1513 who wrote, 'those princes who have done great things … have known how to circumvent the intellect of men by craft'. Part of that craft was lying. Machiavelli argued that rulers should do whatever it takes to retain power, and this could include 'being a great dissembler' (Machiavelli 1513/1977). Politicians lie by omission and by exaggeration, but sometimes they tell outright 'big lies'. This term was introduced by Hitler in *Mein Kampf*.

A big lie 'is a gross distortion or misrepresentation of the truth used as a propaganda technique'. The big lie works, according to Hitler, because the ordinary person knows how bad it feels to tell a small lie so they cannot imagine someone having the 'impudence' to distort the truth so gravely. To work, big lies, according to Hitler, must be able 'to awaken the imagination of the public through an appeal to their feelings'. They are not aimed at our rational selves, but our unconscious and emotional selves (Beattie 2018a).

The first principle behind the big lie is that 'there is always a certain force of credibility; because the broad masses of a nation are always more easily corrupted in the deeper strata of their emotional nature than consciously or voluntarily' (Hitler 1925/2022: 213). In other words, big lies have to be 'credible', although credible is perhaps not exactly the most appropriate word here because they don't necessarily have to be primarily *logically* credible, as examined by our conscious system of thought, but somehow emotionally credible. They have to connect with how we feel, the 'deeper strata' of the human mind, but in Hitler's case they also allowed an emotional outlet for the suppressed emotional feelings (the sense of shock, anger and resentment) of Germany after the First World War. Hitler was thinking here in terms of the depth psychology of Freud and others, that the human mind has layers, the conscious mind and the deeper underlying unconscious mind that can and must be targeted for the purposes of persuasion. The big lie has only to be *credible* to the extent that it doesn't provoke instant conscious rejection. When guided by the right emotions which have been targeted by the lie, the idea itself may start to appear more plausible. Emotions can guide our cognitions in this way, as we have seen, as demonstrated by Damasio (1994) and others. Trump telling us all that immigrants were eating the dogs and cats in Springfield, Ohio is not appealing to our rational system (Where's the evidence? How do they cook them? Are all immigrants at it?). It is providing us with a vivid image, it's an appeal to our feelings, it is trying to impact on our emotional and unconscious system. But we ask ourselves, how could someone have the impudence to tell such a lie? And that, of course, was Hitler's whole point.

These are some of the advantages that political liars have – no constraints on the story or the self-construction, a direct appeal to the emotions without the constraint of truth, an engaging emotional draw. What could be better? And some are very good at it – they suffer little from detection apprehension and feel confident in their ability to succeed. Some politicians are so good at lying, however, you wonder whether they might actually believe their lies or whether it's just down to their personalities. You do search in vain for tell-tale micro-expressions of guilt, shame or sadness, but you find none (Beattie 2016).

As the sociobiologist Robert Trivers has pointed out, lying can give you a clear evolutionary advantage – status, wealth and achievements are important in that great evolutionary battle in the survival of the genes – that's why people lie about them. But Trivers says self-deceit can also be evolutionarily advantageous because if you can convince yourself of something, then it makes you more convincing to others, and therefore more effective. Perhaps Trump managed to convince himself that they really were eating the dogs and cats in Springfield. Maybe he is that self-deluded, as in so many areas (for example about climate change being a Chinese hoax). However, maybe he just thought to himself – plant the image, that is all you need to do for the faithful. MAGA. Attractive fictions might well engage us and sweep us along, but as Shakespeare suggested in *The Merchant of Venice*, and as many people believe—the 'truth will out' eventually. The last few months of the US election campaign suggest, however, that this may not always be true. And then it's not so pleasant for anybody, but especially for the recipient of the lie. For all of us.

DIAGNOSIS?

It is clear that frequent lying without remorse connects to personality disorder. But we have to be cautious about trying to diagnose anyone directly from a distance. Psychiatrists or psychologists cannot on ethical grounds diagnose an individual as being psychopathic or having Narcissistic Personality Disorder without personally assessing them using a standardised clinical interview and procedures (see *The Independent*, 20 February 2017). However, Jeffrey Flier, a former dean of Harvard Medical School, declared on Twitter: 'Narcissistic personality disorder. Trump doesn't just have it, he defines it.'

We know that both psychopathy and narcissism are on a continuum, and Trump may indeed be high on both dimensions, in fact high on all of the dimensions of the Dark Triad, without necessarily hitting the recognised clinical cut-off points. We can't be sure of where exactly he sits without a direct clinical assessment and therefore we cannot apply the label 'psychopath' or 'Narcissistic Personality Disorder'. But we can comment on his public behaviours, we can analyse his lies objectively, we can detail their frequency, their scope, their function, we can explain how these types of lies connect to antisocial personality conditions, we can detail his lack of remorse, guilt, or shame as he lies, we can comment on his absence of emotional empathy (like when he mocked a *New York Times* reporter's disability at a rally in South Carolina in 2015, imitating his movements). These behaviours are all publicly available and open to analysis. We can also consider the negative influence of these behaviours and his false narrative on his millions of followers, as well as on those concerned and anxious about climate.

I was going to leave this section at that, but then on 24 October 2024 (two weeks before the presidential election) an open letter was published in the New York Times signed by 233 mental health professionals with the following warning:

> As mental health professionals, we have an ethical duty to warn the public that Donald Trump is an existential threat to democracy. His symptoms of severe, untreatable personality disorder – malignant narcissism – makes him deceitful, destructive, deluded and dangerous. He is grossly unfit for leadership. Trump exhibits behavior that tracks with the American Psychiatric Association's DSM-V diagnostic criteria for 'narcissistic personality disorder, 'antisocial personality disorder' and 'paranoid personality disorder', all made worse by his intense sadism, which is a symptom of malignant narcissism.
>
> (New York Times, 24 October 2024)

They are clearly aware of the ethical constraint on making a clinical judgement about someone without personally assessing them in a clinical interview, but they say in the letter that they have 'an *overriding* ethical duty to warn the public of the danger this individual poses', and that they have observed thousands of hours of Trump's behaviour and that he clearly meets the criteria for *antisocial personality disorder*. They write:

> Even a non-clinician can see that Trump shows a lifetime pattern of 'failure to conform to social norms and laws', 'repeated lying', 'reckless disregard for the safety of others', 'irritability', 'impulsivity', 'irresponsibility' and 'lack of remorse'. Because of their sadism, malignant narcissists often derive joy from

inflicting suffering on others because they disregard the emotions and well-being of other people – especially their perceived enemies.

(New York Times, 24 October 2024)

They finish with, 'As mental health professionals we feel a desperate duty to warn our fellow citizens of this imminent catastrophic public danger before it's too late.'

This letter broke new ground but, of course, had no effect on the election result.

So, why do some influential people like Donald Trump claim that climate change is fake news? In Donald Trump's case one might suggest that it is the pursuit of power at all costs. He's stoking up populist resentment, encouraging American oil companies that it's business as usual. It's an attack on China ('a Chinese hoax') to bind his supporters together, it's us versus them, to build a sense of solidarity in the in-group, to strengthen his power base. He's reassuring his followers that we're not going to die if we just keep on doing what we do. That's a feelgood message full of optimism and hope that people will like, but unfortunately, it's wrong. He doesn't have to pay any attention to the tricky science because it's not relevant to his message. He plants images in our unconscious and emotional mind, with lies so big that nobody apart from him (and perhaps a few others historically and geopolitically) would have the impudence to tell them, and this impudence undoubtedly links to his lack of empathy and his personality, and it would seem to some an underlying personality disorder. He doesn't seem to care about anybody except himself, that's what those with Narcissistic Personality Disorder are like. His lies about climate change damaged the whole world but not his political ambitions.

And in so doing Donald Trump (and indeed all climate change deniers) contributed enormously to the climate crisis, and undoubtedly to growing climate anxiety. When a person with a recognisable personality disorder argues against your core beliefs about this existential threat we are facing and is supported by millions of people regarding the value of his position and the worthlessness of yours, it is more than concerning. When these same people choose to believe, and vote for, a known and proven liar, it is worse still and then he goes and wins the election. Not a bit of wonder climate anxiety is increasing. Indeed, it makes me anxious just to write these words.

But perhaps the last word here should be left to the philosopher Hannah Arendt (1951/2017) who wrote about the effects of constant political lying on a people:

> the consequence is not that you believe the lies, but rather that nobody believes anything any longer ... And a people that no longer can believe anything cannot make up its mind. It is deprived not only of its capacity to act but also of its capacity to think and to judge. And with such a people you can then do what you please.

FURTHER READING

Arendt, H. (1951/2017) *The Origins of Totalitarianism*. London: Penguin.

Beattie, G. (2024) *Lies, Lying and Liars: A Psychological Analysis*. Routledge: London.

Gounaridis, D. and Newell, J. P. (2024). The social anatomy of climate change denial in the United States. *Scientific Reports* 14: 2097.

Hoffman, A. J. (2015). *How Culture Shapes the Climate Change Debate*. Stanford: Stanford University Press.

Twenge, J. M. and Campbell, K. W. (2009). *The Narcissism Epidemic. Living in the Age of Entitlement*. New York: Atria.

9

HOW CAN PEOPLE STAY HAPPY WITH CLIMATE CHANGE?

I think that this is the question that many who suffer from climate anxiety will want to know the answer to. They have followed the science and understand the implications for the planet and for human life. They know that we need to do something urgently. But many people go about their daily business thinking that they will somehow be immune and that they don't have to worry. How can this be? Here's what my research suggests. Spoiler! It's all in the quick automatic unconscious processing of the human brain.

Information about the science of climate change is readily available, and has been reproduced endlessly in newspapers, television and film. But it is not good news; it is highly threatening and very negative, impacting directly on emotional state and mood (Beattie et al. 2011). But what happens if people avoid seeing this information? What happens if people pay more attention to arguments *against* climate change (which, by definition, present a much rosier picture of both the present and the future – 'there is actually significant doubt about anthropogenic climate change and your current lifestyle is totally acceptable')?

There is evidence from other domains which suggests that some people do have a bias in processing both positive and negative information, and that this bias is linked to 'dispositional optimism'.

DOI: 10.4324/9781032631882-12

Dispositional optimism 'refers to generalized outcome expectancies that good things, rather than bad things, will happen; pessimism refers to the tendency to expect negative outcomes in the future' (Segerstrom et al. 1998). The American psychologist Martin Seligman (2002) has argued that optimists and pessimists differ in terms of one basic psychological feature, namely something he called 'attributional style', which we have considered earlier. Attributional style refers to particular patterns of thinking that we use to reason about why things have occurred, the big and small why questions of everyday life. Why do I feel the way I do? Why do I argue with my partner? Why is my boss so unpleasant with me?

In everyday life when something happens to us, good or bad, we try to work out why it has happened. We have an argument with our partner whilst we're out, we get home and wonder why it happened? Is it because of me, or her, or the situation (supermarket shopping often produces arguments in our relationship I find – too many disagreements about what we want to buy; I hate green tea), or the time of day (shopping after work, never a good idea), or a combination of these various things? We often make attributions quite quickly and spontaneously (sometimes just for the sake of argument – 'You always get tetchy in Tesco!' I shout. I don't actually, but I could), although, on occasion, we may reflect and mull over the possible reasons for hours (and that night is usually spoiled) or worse still for days or longer. That's attributional reasoning in action.

We may decide that the cause of the argument is actually *internal* (it's me!), *stable* (and therefore going to exert its influence over a long and protracted period of time) and *global* (going to affect different things and not just arguments in Tesco). Seligman and his colleagues in 1979 found that if you are the kind of person who naturally adopts such a pattern of attribution (internal, stable and global) for negative events then it tends to make you prone to depression because you're assuming that all bad things are down to you, they're always going to be there and will affect everything you do. You may also tend to explain positive events in terms of external causes (i.e. nothing to do with you). In other words, you take responsibility only for all the bad things that happen to you. This immediate self-blaming is not a healthy state. In terms of therapy, Seligman argued that it helps if you can stop people immediately internalising the causes of negative events and encourage them to recognise instead that events are caused by a combination of internal and external factors or sometimes by external events alone. In other words, to dispute the automatic System 1 judgements. When you fail an exam don't assume that you're stupid (an internal, stable and global attribution). Think more carefully about it, slow the process down. You failed the exam because it was really difficult (i.e. an external attribution – something about the exam rather than something about yourself), not because you're not intelligent or bright enough. It may help, in this particular case, if you remind yourself that many people failed the exam, and that you've passed many exams in the past. Use the data more effectively, don't just jump to a conclusion!

Optimists, on the other hand, according to Seligman, have a very different attributional style, they take credit for all the good things that happen to them, and make external attributions for all the bad things (Seligman 2002). Donald Trump again springs to mind.

However, there may be another psychological factor that also distinguishes optimists and pessimists. Attributional style relates to cognitive reasoning about events, but what about the perception of any such events in the first place. Could this also distinguish optimists and pessimists?

Isaacowitz (2006) has argued that dispositional optimism affects basic perceptual processes and that optimists quite literally look on the bright side of life. He used an eye-tracking procedure to investigate this, tracking individual gaze fixations when participants looked at images of skin cancer compared with line drawings with the same shape as the cancer images, and neutral faces. He selected images of skin cancer because they are clearly 'negative' images, being both unpleasant and graphic. He found that young adults high in dispositional optimism fixated less on these skin-cancer images than their less optimistic peers (Isaacowitz 2006: 68). In other words, Isaacowitz claims that adult gaze preferences 'towards positive and away from negative images suggest that gaze patterns may reflect an underlying motivation to regulate emotions and to feel good' (Isaacowitz 2006: 69).

Luo and Isaacowitz (2007) also reported negative relationships between dispositional optimism and eye gaze to both negative and neutral text about skin cancer. The negative correlation suggests that optimists may read information about a negative topic more quickly than do pessimists. Other research has shown that individual differences in mood are associated with attentional bias to certain stimuli. Individuals suffering from anxiety or depression have attentional biases toward negative information (Mathews and MacLeod 2002). Attentional bias to certain affective stimuli appears to be motivated by the need to self-regulate emotion or maintain one's positive mood state.

Therefore, according to Isaacowitz (2006) and Seligman (2002), optimists have distinct cognitive 'strategies', involving both attention and attributional reasoning, for staying optimistic. At the level of the individual this might be a very good thing because there is evidence that optimists live longer and healthier lives than pessimists (Seligman 2002), and consequently, using a range of techniques, people have been trained to become more optimistic. This is the basis not just of much of cognitive behavioural therapy (CBT) but also the self-help industry.

However, this might be less good for society (or the world) as a whole. Barbara Ehrenreich (2010) has argued that these high levels of optimism have 'undermined preparedness' to deal with real threats, including 9/11, the economic bubble bursting, world terrorism, and you might well add climate change to this list of hers. 'The truth is that Americans had been working hard for decades to school themselves in the techniques of positive thinking, and these included the reflexive capacity for dismissing disturbing news' (Ehrenreich 2010: 10). The economic crisis, she argued, was a case in point ('imagining an invulnerable nation and an ever-booming economy – there was simply no ability or inclination to imagine the worst' (Ehrenreich 2010: 11). Ehrenreich argued that the problem was that 'professional optimists dominated the world of economic commentary' (Ehrenreich 2010: 181) and that some people who had managed to anticipate the forthcoming economic disaster 'had been under pressure over the years to improve their attitude'. This is the downside of optimism, the fact that people may not notice warning signs that are available and that a focus on the negative is actually an important aspect of human survival.

One very significant question is whether optimists may be missing some of the crucial signs of climate change because they are avoiding seeing them. There is some evidence for this. For example, individuals who reported feeling concerned, worried and anxious about climate change were less likely to avoid information about climate change, and more likely to seek out such information (Yang and Kahlor 2012). In our lab, we found that experimental participants fixated more quickly on the carbon footprint of products like low energy light bulbs (designed to be more sustainable), than on the carbon footprint information of products like detergents where the carbon footprint was higher (Beattie et al. 2010).

One important question is: how would level of dispositional optimism relate to attentional focus on more substantive climate change messages? And how might dispositional optimism affect so-called optimism bias, which has been reported frequently in the psychological literature, and may be particularly relevant to climate change?

OPTIMISM BIAS

According to Tali Sharot (2012) around 80% of us suffer from some form of optimism bias in many aspects of our lives – apparently believing that our marriages will work (it's only other marriages that fail, we say), our start-up businesses will succeed, and that we will have a long and fulfilling life compared to everyone else. This sort of unrealistic optimism would seem to be somewhat pervasive, affecting not just our personal relationships but also our attitudes to finance, work and health. For example, adolescent smokers are two and a half times more likely than non-smokers to doubt that they personally will ever die from smoking even if they smoked for thirty or forty years; adult smokers are three times more likely to believe this (Arnett 2000). When it comes to smoking or climate change, this optimism bias can have deadly consequences.

Optimism bias has been found across a range of environmental issues (Gifford et al. 2009), as well as in estimates of the risk of health damage from specific environmental hazards, like water pollution (Pahl et al. 2005), and climate change (Gifford 2011). A large eighteen-nation survey demonstrated that individuals believe that across a number of environmental issues they are safer than others living elsewhere and that they are safer than future generations – in other words, they show both a spatial and a temporal bias.

Optimism bias appears to be associated with specific cognitive biases in processing relevant information. One study in behavioural neuroscience used Functional Magnetic Resonance Imaging (FMRI) to measure brain activity as participants estimated their probability of experiencing a range of negative life events, including things like Alzheimer's and burglary (Sharot 2012). After each individual trial, participants were presented with the average probability of that event occurring to someone like him or herself. The researchers found that their participants were significantly more likely to change their estimate only if the new information was better than they had originally anticipated. This bias was reflected in their FMRI data in that optimism was related to a reduced level of neural coding of more negative than anticipated information about the future in a critical region of the frontal cortex (right inferior prefrontal gyrus). They also found that those participants highest in dispositional optimism were significantly worse at tracking this new negative information in this region, compared to those who were lower in dispositional optimism. In other words, the optimism bias derives partly from a failure to learn systematically from new undesirable information and this bias was most pronounced with those highest in dispositional optimism.

Optimism may be highly advantageous for the individual, as Seligman has consistently argued, because it has significant effects on both mental and physical health (Seligman 2002) and was selected for during evolution (Mosing et al. 2009). Optimists live significantly longer and are much less likely to die from cardiac arrest (Scheier et al. 1989); optimism also increases the survival time after a diagnosis of cancer (Schulz 1996). It does this by reducing stress and anxiety about the future, and optimists consequently have better immune functioning (Segerstrom et al. 1998). Belief in a positive future also encourages individuals (in *some* domains, particularly those that they have some control over) to behave in ways that can actually contribute to this positive future, thus becoming a self-fulfilling prophecy (Sharot 2012). Although underestimating future negative life events can reduce our stress level and add to our longevity, sometimes negative events really do need to be considered. Hence, optimism bias can have very significant deleterious consequences, particularly regarding the discounting of serious risk.

Optimism bias is particularly relevant to issues to do with climate change. If we underestimate the probability of the negative effects of climate change happening to us, we may be much less likely to engage in mitigation behaviour, or sacrifice many of the things we currently value (foreign holidays, big cars, and high carbon lifestyles) to reduce the risks associated with climate change. But how might we gain insights into cognitive biases in the area of climate change? We cannot present participants with the actual outcome data (as they did in the Sharot study) to see how they update their own estimates in the light of this, because the catastrophic consequences of climate change are still largely in the future. However, could the research on biased patterns of attention provide us with any new insights here? After all, we are constantly being presented with articles documenting the scientific consensus on the likely effects of climate change, but could there be cognitive biases in how we attend to these messages? Does the evidence of differential focus on positive images by optimists in the Isaacowitz study have any relevance for how individuals attend to more serious substantive messages, as opposed to drawings and images?

ROSE-TINTED GLASSES

We decided to test this (Beattie et al. 2017). We used three articles about climate change – the first about climate change in general, the second about climate change and its relation to flooding in the UK, and the third about climate change and its consequences for food scarcity and violent conflict. Each climate change article contained three arguments for climate change ('for') and three arguments against climate change ('against'). 'For' arguments were that climate change is real, human activity is the cause of both climate change generally and flooding in the UK, and predictions that climate change will cause food scarcity and conflict. 'Against' arguments were that climate change is not occurring or is exaggerated, that it is not caused by human activity, that flooding in the UK is not caused by climate change, and that there is no link between climate and food scarcity and conflict. All arguments were drawn from print and electronic media (e.g. *The Guardian*, BBC *News* website, etc.) and online blogs. 'For' and 'against' arguments were edited such that they were of similar word count and frequency.

We also measured dispositional optimism using the 10-item Life Orientation Test, which asks participants to respond to a series of simple statements like 'In uncertain times I usually expect the best'. A high score on this (e.g. 'I agree a lot') is taken to indicate a very high degree of optimism. We then split our participants into two groups, optimists and non-optimists, based on their median score.

The individual scan paths showing individual fixations (represented by the circles) and the eye movements (the saccades) of two participants are shown in Figure 9.1 and a 'hotspot' analysis in Figure 9.2.

The results of this study were very revealing. We found no significant relationship between level of dispositional optimism and the number of fixations on arguments either 'for' or 'against' climate change. However, there was a significant relationship between level of dispositional optimism and average fixation *duration* on 'for' arguments only. Optimism level was also significantly negatively correlated with average dwell time on both 'for' and 'against' arguments.

Previous IPCC reports on climate impact have been plagued by errors that have damaged the body's credibility. Most famously, in the 2007 report, it said that glaciers in the Himalayas could disappear by 2035, a claim it has since withdrawn. One reason for errors in the IPCC reports could be the over-reliance on computer models of predicted data, rather than on physical science.

The recent IPCC report raised the threat of climate change to a whole new level - based on new scientific evidence - warning of sweeping consequences to life and livelihood. The report concluded climate change is already having detrimental effects – melting sea ice in the Arctic, killing off coral reefs in the oceans, and leading to heat waves, heavy rains and mega-disasters. And the worst was yet to come.

Previous IPCC reports on climate impact have been plagued by errors that have damaged the body's credibility. Most famously, in the 2007 report, it said that glaciers in the Himalayas could disappear by 2035, a claim it has since withdrawn. One reason for errors in the IPCC reports could be the over-reliance on computer models of predicted data, rather than on physical science.

The recent IPCC report raised the threat of climate change to a whole new level - based on new scientific evidence - warning of sweeping consequences to life and livelihood. The report concluded climate change is already having detrimental effects – melting sea ice in the Arctic, killing off coral reefs in the oceans, and leading to heat waves, heavy rains and mega-disasters. And the worst was yet to come.

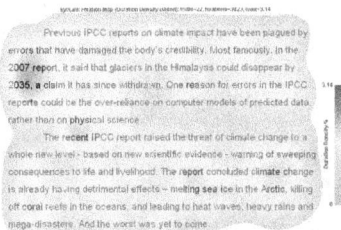

(a) Optimist   (b) Non-optimist

Figure 9.1. An individual scan path of (a) an optimist and (b) a non-optimist, as they read arguments for or against climate change. Circles represent individual fixations on words, with larger circles representing longer fixation durations. Lines between circles represent saccadic eye movement behaviour. In this example, the first paragraph is an argument against climate change and the second paragraph is the argument for climate change.
(from *Beattie et al.* 2017)

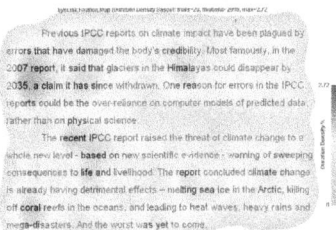

(a) Optimists   (b) Non-optimists

Figure 9.2. A hotspot analysis of eye gaze fixations of a group of optimists and non-optimists reading arguments for or against climate change. Darker areas represent longer dwell times at fixated locations.
(from *Beattie et al.* 2017)

Thus, higher levels of dispositional optimism are associated with less time spent attending to the content of the climate change articles irrespective of particular argument (either 'for' or 'against'), and shorter periods of time fixating on arguments 'for' climate change. In other words, optimists are turned off by any messages about climate change (either positive or negative), but they also spend less time fixating on messages arguing for climate change.

But how did this on-line processing affect what people remembered about the messages, given that in each article there were arguments both for and against? We tested this by asking our participants to summarise the articles. We measured overall level of recall, which did not significantly differ, but we also analysed how each recalled article was framed. We employed three broad categories for coding how these recalled accounts were framed.

'For': the account was framed as being primarily about the evidence for climate change (and its general effects or specific effects on flooding, food scarcity and conflict, etc.) and the role of human activity in this.

'Against': the account was framed primarily in terms of there not really being a strong link between human activity and climate change (or its specific effects), or doubts about the very existence of climate change.

Debate: the account was framed as primarily being a debate between two opposing positions.

We found that non-optimists, who had longer fixations on the arguments for climate change, were most likely to frame their recall in terms of the arguments for climate change ('this article is about global warming and how 95% of it is due to human activity); 66.7% of their recalls were framed in this way. The optimists, on the other hand, who fixated significantly less on arguments for climate change, were more likely to frame their recall in terms of a debate between two opposing positions ('it's about climate change, about trying to understand what's happening with the weather and there are different points of view'); 66.7% of their recalls were framed as a debate. There were few summaries of the content framed in terms of the arguments against climate change for either group (only 5% of the total).

Dispositional optimism thus seems to affect on-line processing of climate change messages and the framing of how these messages are recalled. We then decided to consider the relationship between level of dispositional optimism and the extent of optimism bias. With a new set of participants, we again measured level of dispositional optimism and devised a simple questionnaire to measure optimism bias. It consisted of three broad questions:

1. What is the probability of you personally being affected by climate change?
2. What proportion of people (living today) will be affected by climate change?
3. What proportion of future generations will be affected by climate change?

Participants had to write a number between 0 and 100% in response to each of the questions. Each question had seven additional questions asking participants to rate (in the case of question 1): the probability of them being personally affected by severe drought/ severe flooding/major threats to infrastructure/food shortages/ major conflict/heat-related increased mortality and major disruption to your life. In the case of the other questions, they had to rate the proportion of people living today (question 2) being affected by each of these, and then the proportion of future generations (question 3) being affected by them. There were thus twenty-four questions in all to assess possible optimism bias. We split our participants into three groups – 'optimists' (optimism score 18–23), 'medium-level optimists' (optimism score 15–17) and 'non-optimists' (optimism score 8–14).

We found that optimism bias is significantly affected by underlying level of dispositional optimism – for example, optimists in our study reckoned that they had a 36.5% probability of being personally affected by climate change, whereas they thought that other people had a 52.8% probability of being affected, and that 76.4% of future generations would be affected. For non-optimists, the figures were higher throughout – they reckoned that they had a 56.8% probability of being personally affected by climate change, and that 68.5% of other people would be affected, and 84.1% of future generations would be affected. Even non-optimists have some degree of optimism bias – they thought that they would be less likely to be personally affected by climate change than other people elsewhere and in the future. But the optimists in this sample were particularly blasé (about a one in three chance, they reckoned, that they would be personally affected), which is perhaps worrying given that there is a major self-help industry devoted to training people to be *even more optimistic*. However, sometimes a little *realism* about events and their causes in everyday life, as a guide to future appropriate actions, is very important.

Of course, this research into cognitive biases, and particularly optimism bias, has a number of potential general implications. We cannot assume that all members of the public are attending to messages about climate change in quite the same way (regardless of how credible the source is). The underlying messages may not be getting through because of an inherent cognitive bias designed to sustain emotional state, and, of course, a very robust bias shaped by evolution. The research perhaps suggests that we should pay some regard to this bias in designing our communicational strategies about climate change. It may well not be enough simply to publicise the scientific evidence about climate change without framing some of it in a more optimistic way to highlight the positive aspects of mitigation strategies. A more positive overall frame about possible positive solutions should increase both feelings of self-efficacy and visual attention to the underlying message. Without this, we have the grave danger that many will selectively attend to the information and ultimately show little behavioural adaptation or indeed concern.

There is another important consideration here that is related to this. For the past few decades, we have been striving to increase optimism in society because of its health benefits (through both positive psychology and a cultural emphasis on 'the power of positive thinking'). Some have argued that we have produced a profound socio-psychological change, especially in Western societies, with unrealistic expectations about the future (Ehrenreich 2010). They have also argued that it has actually 'undermined preparedness' to deal with real threats like global terrorism, financial bubbles, or climate change, with the public having 'no ability or inclination to imagine the worst'. Optimism can be a positive thing, but it has its limits. Over-optimism can be very damaging. Perhaps, it is time to re-evaluate this overarching cultural focus and consider new ways to get the public to imagine the worst. However, this is unlikely to occur successfully on its own, unless we also spell out ways that we can mitigate these effects.

So, what about the question I posed at the start of this chapter – how do some people stay so optimistic about climate change? The answer would seem to be that their attentional focus shifts when they read climate change messages, and they unconsciously focus on the good news sections. This is linked to underlying personality. Dispositional optimists are more inclined to do so. In other words, their brain guides them away from the bad information; they avoid reading it and they feel safer. They remember the articles differently; they feel less threatened by the climate change stories. System 1 responses are at the heart of it, automatic and non-conscious, the workaholic System 1, and these optimists manage to maintain a positive mood amid this existential threat and hang on to their positive and jolly outlook on the world. But of course, this is very dangerous for all of us, them included.

FURTHER READING

Beattie, G., Marselle, M., McGuire, L., and Litchfield, D. (2017). Staying over-optimistic about the future: Uncovering attentional biases to climate change messages. *Semiotica* 218: 22–64.

Ehrenreich, B. (2010) *Smile or Die: How Positive Thinking Fooled America and the World.* London: Granta.

Isaacowitz, D. M. (2006). Motivated gaze. The view from the gazer. *Current Directions in Psychological Science* 15: 68–72.

Seligman, M. (2002) *Authentic Happiness. Using the New Positive Psychology to Realize Your Potential for Lasting Fulfilment.* New York: Free Press.

Sharot, T. (2012) *The Optimism Bias. Why We're Wired to Look on the Bright Side.* London: Robinson Books.

10

IS THERE A STIGMA SURROUNDING CLIMATE ANXIETY?

I very rarely mention my climate anxiety or whatever it is (it sometimes feels very vague inside, like some form of psychical disturbance). I don't want to be perceived as weak or neurotic. I'm a scientist for goodness sake! This book is my confession. Perhaps, I'm just a coward. Or a realist? How do other people think about climate anxiety, where 'other people' here means those who have never experienced it? How do they think about those who do? Is there a stigma surrounding it, as there was with shell shock at the beginning of the last century? Is it perceived as a device for malingerers, a strategy to get special treatment, a trick to elicit compassion, something to occupy the minds of snowflakes? This chapter looks at talk about climate anxiety and some subtle and some less subtle social constructions of victims and victimhood through discourse.

If there is a stigma around climate anxiety, this can only make people who suffer from climate anxiety feel worse and quite possibly more anxious. To be dismissed as 'neurotic' when you suffer from severe anxiety about what is happening to the climate is not just unhelpful, it can be very damaging. This is not just some whim of mine. In Norway, Gregersen and his colleagues (2024) asked members of the Norwegian public what they thought when

DOI: 10.4324/9781032631882-13

they heard or read the term 'climate anxiety'. The majority of the respondents (52%) viewed it neutrally as a synonym (more or less) for worry, but a significant proportion (27%) viewed climate anxiety very negatively, as *unfounded, irrational,* or *excessive*. The researchers also found stronger support amongst the public for politicians taking young people's climate concern or climate *worry* into account when designing new climate policy compared with young people's *climate anxiety*. They concluded that referring to climate anxiety 'may provoke reactance among some audiences'. *Reactance* is a defence mechanism to ward off threat by assuming that the other person is trying to manipulate you. It is a form of denial. A significant proportion of the public clearly don't like the term and quite possibly also the people who suffer from this unfounded and excessive condition.

But as always with this sort of research, you want to know more. How do they see people who suffer from climate anxiety? How do they understand the condition? What does it entail? In the summer of 2024, I wanted to gain some insight into the issue by just listening to what people were saying to me about this, university-type people (because that's where I hang out), educated people. This is how climate anxiety is socially constructed in talk; this is how stigma around it is negotiated and managed.

THE SIGN IN THE KITCHEN

This small dark room of mine in the university did not look much like an office, it looked more like a cave, or a sanctuary. As if I might be hiding away from malignant forces, perhaps even from the weather itself, whispering to me that the climate was changing. It was summer but it was cool in the room, even slightly chilly, and many would comment on this as they came into the room. It acted as a sort of stimulus to conversation about temperature, heat, my house plants, the weather, and sometimes even the climate. It was like a natural experiment.

'It's lovely and cool in here,' said one student as she came in, 'not like outside, it's really warm today. It's a real hot spell. Maybe, it's climate change, you never know.' And she laughed at her own suggestion.

We both more or less simultaneously glanced out through the dense foliage just outside my window as if to reaffirm the message, to see a land baking, well not so much baking but warming up slightly – it was after all the UK. But, of course, the conversation gave me that opportunity to ask about the weather more generally (that great British staple) and then climate change.

'It is warm today … but I was just wondering,' I said gently, 'do students around here worry about climate change much – do many suffer from [there was a slight polite pause] climate anxiety?'

It was both a general question and deviously specific. 'Around here' – where's here? The North West of England (the region near Manchester and Liverpool), the university, where I'm currently sitting?

'Some do, but for me, it's yes and no', she said in a noncommittal sort of way. 'Yesterday I woke up and looked out my window and I thought great it's sunny again. But after the third or fourth day I thought perhaps there is something going on. After all, it's the North West of England. Blackpool isn't that far away and the weather there has always been horrible. But I'm not that concerned because I don't think I'm a worrier.'

'What do you mean?' I asked politely. 'Not a worrier.'

'Well, I'm a student, so I do worry about some things – I worry about my exams, my workload, my flatmates, my boyfriend, especially at the moment, money, debt, not about the sun shining. It's not that important in the grand scheme of things.'

She looked at me for reassurance. I was surprised that she hadn't mentioned the wars in Ukraine or the Middle East, but perhaps she took those for granted. That would have been a more serious conversation.

'But my flatmate is different to me', she continued. 'She's a real worrier and this summer in particular she's really worried about climate change. I would say she has genuine concerns as she's been reading more and more about climate change, and she says it's all getting worse. But it's also her personality. She's quite up and down really, although the truth is most days she's down.'

My ears pricked up when she mentioned her friend's 'personality'; I suspect that's how many think about climate anxiety, it's something that afflicts neurotic people.

I asked her whether her flatmate was more worried about the climate than about her exams. It felt like a guilty sort of question.

'Not *more* worried,' she replied, 'but it's just one other thing for her to worry about. She seems to have a list of things to worry about, but sometimes things change in position. These are things that keep her awake at night. She is a very bad sleeper; from what I can see. I think climate change is about number two or three at the moment. She gets on to that one about three in the morning after she's worked through one or two of the others.'

I asked my student what her friend's number one worry might be at the present time.

'It's hard to say but I think it might be people. She doesn't have much faith in them. She's told me that it's people like me that worry her the most because we're all blind. We're intelligent but we can't see what's going on. She says that I, and so many people like me, don't take climate change seriously enough. She's even started marking up articles for me to read as she knows that I can't be bothered to read the whole thing. She tells me just to read the marked sections. I said to her that I've got enough articles to read for my exams and I don't really need any others. But she just gets upset. I get upset too; I don't want her to think that I'm heartless.'

I paused briefly. It was only meant to be a casual conversation, but my student said that she wanted to talk more about her friend.

'She's a big fan of Greta Thunberg – you don't hear so much about her these days unless you really care about climate change but she's still a big influence although she's moved onto other things. My flatmate has written 'OUR HOUSE IS ON FIRE' in bright pink felt tip pen and pinned it up in our kitchen. A couple of my other flatmates think that it's a bit inappropriate and have told her to take it down. My friend Olivia says that it's putting her off cooking in the flat because it reminds her of the fire risk from the cooker. The cooker's a bit old and very dodgy. Olivia says that she has started eating more takeaways because of it, and that's not great for her diet. The kitchen also stinks because she never throws out the food cartons from the carry outs. And Olivia is putting on weight and blaming it on Greta Thunberg. It's all got a bit stressful.'

I asked whether she ever thought her climate-anxious flatmate was right because if our house (or our planet) is on fire and we're all just sitting there watching TikTok videos or whatever then perhaps we should all be panicking. Could climate anxiety be a really good thing and a driver of change for us, something to get us all to act?

'I'm not sure about that,' the student replied, she was looking very reflective, 'because my friend does less and less with regard to sustainability in her own life. At least, that's what I've observed. She doesn't even recycle anymore. She just says what's the point because of people like me – she's trying to make me feel worse and worse. I'm not sure she'll even pass her exams this year. She stays up all night worrying and sleeps all day. She's actually seeing a counsellor at the moment and the university have been great, really supportive, but she tells me that she just talks to her counsellor about all her other anxieties; the session is always finished before she gets round to climate change anxiety. That's what she says anyway.'

I nodded, but I was not quite sure what I was nodding at.

She continued, 'And my flatmate always says that she wouldn't be comfortable talking about climate change and the anxiety associated with it because she says you don't know how the counsellor herself thinks about this. She knows that people have different views on this and that some people might think you're a crackpot if you go on about it. She says it's better for her to stick to talking to her counsellor about her exams and her boyfriend troubles and her difficult and aggressive flatmates and her finances, all the sorts of things that the counsellor might actually understand. I've thought about reading the articles for her so that we could talk about it but sometimes I think that might make her worse. If I agree with everything she believes and I tell her that I think that the planet is on fire, then it could tip her over the edge. I think whatever I do it's going to make her worse about herself and that's making me a bit anxious. I'm not sure if that counts as climate change anxiety or not but it's certainly impacting on me and it's all related to climate change.'

'Oh', I said. 'I'd never really thought of that.' And she got her essay out to signal that this conversation about climate anxiety was now closed.

This conversation did get me thinking though about how broad or specific the concept of climate anxiety might be, it seems to radiate outwards, as well as this idea that only certain kinds of people might be affected. There did seem to be a degree of stigma. So, I went in search of a colleague to invite her to my office for a chat. She was likely to have a different perspective on climate anxiety, she was a climate change researcher, maybe even an expert in the area. She has been working in local schools for a number of years trying to change children's attitudes to sustainability to get them to act more sustainably and not just say that they care about the planet. She's on the frontline of that great divide between mere pro-environmental statements and actual sustainable action. She's in a very different position from the student but had a similar response as she came into my office.

'Gosh, it's chilly in here,' she said, 'not like my office, it's boiling there, the sun shines right through the big windows.'

The truth is I envied her big windows, but I suppose there's an advantage to having an office where trees block out the light. I mentioned my last conversation briefly and told her how interesting I'd found it and asked her when was the last time she suffered from any degree of climate anxiety. There's strong evidence to suggest that certain ages, demographics and professions are susceptible to climate anxiety. Climate change 'experts' are a group that are particularly affected because of what they know about the precarious state of our planet.

'I don't suffer from climate anxiety as such,' she said, 'and if I was asked why I would say it's probably because I try to live sustainably and so do many of my friends and family. If I lived in Dubai or somewhere like that – someplace founded on this ridiculously high carbon lifestyle, then I might be much more worried, but I feel that climate change is a problem that people know about and they're trying to do something meaningful. We have to help them. I suppose that puts me slightly at ease.'

'But do you never feel anxious in that boiling office of yours when it's very hot like today?'

'It's just the weather,' she said, 'the weather goes up and down. I know that sometimes when the weather's very warm people think slightly more about climate change and when the weather's bad, they seem to dismiss it, but that's them. I take the weather with a pinch of salt, if that makes sense. I did say that I don't experience climate anxiety directly but that's not to say that I don't feel anxious about some of the things associated with it. Let me tell you a funny story, I had to take a bottle of water to work a few days ago. It was given to me by my partner after a running race, the problem is he doesn't drink much water, so he gave it to me. Now I drink a lot of water, but I would never buy a bottle of water, I mean a plastic bottle full of water, I always fill something up from the tap. But I didn't want to throw this bottle of water out, think of the waste, so I brought it to work with me and then I thought to myself what happens if someone sees me drinking from this. The more I thought about that the more anxious I got. Would they think I was some kind of hypocrite, and would I think differently about myself? I try to live sustainably. So, I put the bottle in my bag, I sneaked outside and sat on a bench near the library and started to drink from it. But I swear if anyone had seen me and noticed my body language then I honestly think they would have assumed I was some sort of alcoholic sneaking out for a drink of alcohol. I looked so furtive! I was so concerned about anybody seeing me. To be truthful I felt a high degree of anxiety about that so that was anxiety associated with climate change but not climate anxiety as it's normally talked about. When I look back on it now, I still feel a bit anxious about it. I'm embarrassed just talking about it like it's some awful confession.'

This was another interesting personal account because it showed how climate change and climate anxiety can impact on us in different ways. It can put us under different sorts of pressures. It's not just anxiety about the earth burning up, or our house being on fire, there's also a different sort of anxiety about whether or not you're doing the right thing or, perhaps in this case, whether you're *perceived* as doing the right thing.

My colleague wanted to clarify her account.

'I feel that I should say that if I hadn't been on campus that day, I wouldn't have felt so anxious about drinking water from that plastic bottle, but I still wouldn't have been happy about it because I think it would reflect on the kind of person I am. But then again, what should I have done with the full bottle of water? Just throw it away? Give it to someone who didn't care? Perhaps, I'm a little over-concerned about what other people think of me, but I clearly felt guilty about that, and that guilt was associated with a degree of anxiety.'

So, the conclusion would seem to be here that some climate anxiety can result from social pressures associated with climate change but it's clearly not what you normally think of when you think about climate anxiety. But I wanted to know how severe this anxiety was compared to other things. The list in the night which keeps people awake.

'Well, I suppose if I was to keep a diary and write down all the things that have made me really anxious in the past few weeks it would be on the list, but it wouldn't be at the top. What was at the top was my experience the other day when I was trying to put some empty bottles in the bin for recycling. When I lifted the lid off the bin store which lifts the bin lid itself, I saw a rat running around the top of the rim. I shrieked and ran straight inside. I just didn't know what to do. This was the bin for recycling glass, and I had a number of bottles to bring out, but I didn't want to go anywhere near that particular bin. Later that night I went back out and dumped the bottles in the bin for general waste. I knew I'd feel bad about that, I'd feel guilty and indeed anxious, but it was better than seeing that bloody rat again. My anxiety about recycling and climate change was less than my anxiety about facing that rat. I know that the rat's still out there somewhere, and every time I think about the bin, I get anxious.

My anxiety associated with climate change is in a kind of hierarchy of fear. Right at the top at the moment is the rat, and I suspect that's got something to do with human evolution. Things like rats have been about for a long time, they're very real and they're an obvious threat because of disease. They make me shudder. Climate change is very real but it's not going to kill me or injure me today where I live in the UK. Things might be different if I was living in Australia or some island in the South Pacific but I'm afraid the rat takes precedence in my particular hierarchy of fear. I don't go anywhere near the bin now, even after the bin was emptied. Who said fear had to be rational?'

So, it would seem that climate change may be one significant source of anxiety for some or many people, but it still has to be situated amongst other anxieties, which can push us in different directions.

NOT YOU MY FRIEND, SURELY?

My third conversation about climate anxiety was in Sweden a week or so later. I was sitting in a cosy restaurant waiting for an old friend to come along; I had told him that I was visiting, and he had suggested going for dinner. He was a very distinguished professor of logic and communication at a major Swedish university. We hadn't met for many years, but I had always admired his cleverness, his directness and his ability to get to the point, to deconstruct arguments and sometimes people in the process. He was, he liked to point out, a logician after all, he didn't like fuzziness or contradictions. He hadn't changed that much in appearance and was still as direct as ever. He is very tall (I'm not) and has a presence; he clenched and shook my hand firmly, taking the upper hand.

I started to look at the menu, he pointed to the *inglad sill* (the pickled herring). 'You must want something traditional', he said, and ordered it for me. He asked what I was currently working on. I explained that I was working on a book on climate anxiety. He made a face. It looked like a face of disapproval.

I asked him what he thought about climate anxiety. He laughed. 'You mean Greta? Our most famous export after Volvo and IKEA.' And he laughed at his own joke.

'Some people, as we both know, can be a little [there was a very long pause] ... neurotic, can we say.' The pause wasn't occasioned by a search for the right word, it was for dramatic effect.

I interrupted him (something I don't think he has ever really got used to at any junction in his life) merely pointing out that climate change is this existential threat, the science is clear, and the consensus on the science is like nothing we've seen before, absolutely compelling. So, what can we do when we've been told that we're all facing this kind of threat? Isn't some kind of generalised anxiety the only *rational* answer to the problem? Isn't that the kind of thing we should all feel when we're locked in a house that is on fire with no one listening?

Maybe it was my use of the word 'rational' that angered him. He picked at his herring with his fork. The picking resembled a sophisticated sort of stabbing movement. I was waiting for his response, but he simply turned it around. 'Let me ask you something', he said mid-stab, 'do *you* suffer from climate anxiety?'

'Ah', I said, but it was a slightly long 'aaah', almost imperceptibly too long. My pause made him smile broadly with a slight guffaw of a laugh. I wanted to consider the implications of what I might say (you have to be careful when you talk to a logician, that's what I've always found), my filled pause, my 'ah', was there to shorten the silence as I thought of my reply, to save my embarrassment. I wanted to find exactly the right word. I was almost about to say 'concerned'.

'Aaah', he said back, smiling at his own cleverness. 'So, what does this "aaah" mean?'

'*Well*', I said.

'Yet another filler!' He jumped in immediately; he was smiling even more now. Perhaps that's how logic works, you ask someone a simple question (so simple that the person needs to think about it), wait for any momentary silence, jump in and watch them squirm. You keep them squirming by applying the pressure. You are able to recognise parts of speech as they are produced, paralanguage and all. It all seemed slightly more emotional than rational. Any slight pause seems to be interpreted as a sign of weakness, a signal to attack.

'I think I do get *a bit* anxious about climate change,' I said, '*sometimes* but not all the time [I didn't want to appear weak], but I believe I'm doing all I can to make people aware of it through psychology, and that reassures me a bit.'

But I noticed that he had that look on his face, a slight smirk, which I hadn't seen for many years. I had seen some very famous psychologists, sociologists and linguists many years ago confronted with that look in Cambridge and in London and just wither away because of that expression and the questions that accompanied it. Always right to the point. He encouraged me to come out of my shell back then, to ask more questions, to be more direct. He had seen something in me and encouraged me. I was always very grateful for that, and I had changed. But now he was using it on me.

'But does your psychology *ever* work?' he asked.

'Ever', he repeated, with that smirk again, as if he was expecting some great confession. 'And how would you know if it worked? How would you test it? Are you perhaps just fooling yourself?'

There was a slight pause, so he continued.

'And you're saying that's why you aren't feeling more anxious. I think that could do with a bit more analysis. Maybe it's just a bit of post-hoc rationalisation. Maybe inside you really feel that climate anxiety is just a form of neurosis, and that Greta is the neurotic-in-chief.'

That smirk was starting to annoy me.

'And if you are going to work on climate anxiety or climate neurosis, or whatever you're going to call it, then you need to start with some definitions and some basic propositions, outline what we know, and don't forget we need some clarity here.'

'*Charity?*' I asked. It was his Swedish accent, I wanted to make sure that I'd heard him correctly.

'No, of course not, the opposite of charity. *Clarity!*'

Another pause opened up. This was starting to not feel like a conversation between old friends. Climate anxiety had done that.

It divides people; it seems to make people hostile, but why? I picked at my herring (he had told me to order it, and I didn't like it, I was starting to resent it) and kept my head down. 'Eat up!' he said. 'Viking food.'

I thought carefully about what I was going to say and then changed my mind. I didn't want an argument.

'I'll tell you what. I'll get back to you when I finish my research, let's leave it at that', and I stabbed at my pickled herring to break it up to make it look as if I might be eating it, and I changed the subject completely. Then we reminisced about old times, the old times that we once knew before all of these global crises hit, the good old days which were certainly better than this. Not just faced with waves of existential crises and with a barrier between us. We weren't even on the same team anymore.

My own anxiety level was rising. This professor sitting opposite me was a very clever man whom I had admired so much over the years when he talked about language and logic in numerous academic discussions, but he didn't seem to get it with climate change or even attempt to understand climate anxiety. Logic is sometimes a weapon wielded in debate, not an even-handed emotion-free calculus to test valid arguments. I think that a professor of logic had just demonstrated this to me.

Climate anxiety is a topic marked by stigma and division. There are two sides here with clear social identities constructed in front of our very eyes and strong emotions for and against, strong positive feelings for the in-group, and stigmatisation of the out-group. Two sides, from my friend's point of view, rational versus emotional ('the scientists will find a solution – like carbon capture' was his educated opinion), balanced versus unbalanced, resilient versus neurotic. He clearly felt that he belonged in the first of each of these paired categories. He wasn't sure where I belonged and perhaps that uncertainty angered him, or maybe he thought that I'd drifted to the Greta camp and was too weak to admit it. Could that be why there were so many mentions of Vikings and Viking food? Was he suggesting that I should man-up – be more Viking?!

It is interesting that climate change and climate anxiety generate so much division and so much heat in discussion. The science is compelling so why don't clever people like my friend get the message. And why are they seemingly getting hostile about those who do get the message? It also occurred to me in that restaurant that logic and the rational mind sometimes follow our emotional responses rather than guiding them. We feel certain things and then we introduce our logic and rationality to defend our feelings and attack the other position, as Antonio Damasio (1994) had suggested. Damasio had found that activation of the emotional system precedes activation of any conceptual or reasoning system and that the two systems are separate. Damasio and his colleagues demonstrated that, 'in normal individuals, non-conscious biases guide behaviour before conscious knowledge does. Without the help of such biases, overt knowledge may be insufficient to ensure advantageous behaviour.' In normal people, activation of the emotional system precedes activation of the conceptual system, and Damasio's work suggests that the neural connection between these two systems is located in the ventromedial prefrontal cortex.

So, did my friend and great logician not get the appropriate emotional response to climate change? And is that why he couldn't understand, or be prepared to contemplate, climate anxiety? And if he didn't get that emotional response, why not? It's those feelings all over again, pushing us one way or another. And how important are feelings to climate change mitigation? Do we have to move beyond knowledge and logic? I was trying to understand those who suffer most acutely from climate anxiety and attempt to work out what brought them to this and what if anything psychology could or should do to alleviate their 'symptoms', the symptoms perhaps of a dying world, perhaps the symptoms we should all be showing, but aren't.

FURTHER READING

Beattie, G. and Ellis, A. (2017). *The Psychology of Language and Communication*. London: Routledge.

Part 4

_____

THINKING OF THE FUTURE

11

NEW HORIZONS?

Many people suffer from significant psychological distress because of climate change but there is still disagreement on how we should define and measure 'climate anxiety'. How does it fit with clinically recognised types of anxiety? The simple answer would seem to be that it doesn't fit, at least not easily. It is unique in so many ways and that brings special scientific challenges for those studying it, and personal challenges for those who suffer from it. And it's emerging in a context of assumptions, false beliefs and downright hostility, in my view somewhat reminiscent of shell shock a century ago. It's not currently recognised as a clinical disorder. I have considered the arguments as to whether this needs to change. But it is a dilemma for many, including myself. I'm on that very uncomfortable fence. But in the meantime, is there anything we can do to help ourselves, to allow our anxiety to be more adaptive and attuned to the climate crisis we all face? Are there any new horizons?

I remember 2007 well and the excitement of Tesco launching its carbon labelling scheme. There was such a feeling of optimism that day, such a buzz about this new green revolution. The Prime Minister of the UK was beaming, the CEO of Tesco was beaming, I might well have been beaming (but I can't be sure). A leading multinational was being encouraged (and supported) by the government to engineer a green revolution driven by consumer demand – ordinary members of the public would see the carbon footprint of each product

DOI: 10.4324/9781032631882-15

and choose wisely. High carbon products would be driven from the shelves by consumer choice. All of Tesco's consumer surveys said that the time was right. The price of the green products was carefully considered and kept comparable. This initiative was going to have a knock-on effect on all other retailers and the industries that manufactured the products. There was going to be a massive ripple effect. There was going to be a revolution! What could possibly go wrong?

… Everything, it seems. It just didn't work. Consumers just didn't look at the carbon labels in the way anticipated. They didn't *feel* strongly enough about them to see the label in that five- to seven-second window of individual product selection in supermarket shopping. Their actions didn't match their words. I was dismayed, Tesco was dismayed, for all I know the Prime Minister was dismayed. The fall-out was enormous.

Large international companies started to doubt whether the public really did care about the carbon implications of products in the way they said; they started focusing on other things. The public said that large companies weren't doing enough; they didn't care. And everyone blamed the politicians. That day in 2007 at the Royal Society in London, everything seemed so *integrated*, so joined up, so together, so optimistic; after that, everything seemed so fractured.

I think my own first inklings of climate anxiety happened soon after that. Negative thoughts to begin with, exacerbated by noting these discrepancies between words and actions with regard to climate change, everywhere I went, including in myself (I have to say) as I became more and more dispirited. I went to climate change meetings at UNESCO in Paris and I watched these enormous black limousines sitting in long rows waiting to pick up delegates (not from our group I should add!) to take them a mile or so to their fancy Parisian hotels. I read about Al Gore's lifestyle, somewhat at odds with what he was preaching in *An Inconvenient Truth*, and all the other celebrity endorsers of the climate change message. I watched as Tesco halted its carbon labelling scheme. I got depressed.

But I felt that I needed to try to learn a bit more about why people don't act in accordance with what they say, to uncover implicit

attitudes and systems of cognition, to rethink the nature and importance of feelings in decision-making, to expose cognitive biases.
At the back of my mind was this monumental worry that people
might be saying the right things (even politicians) but their actions
just don't match. Greta Thunberg, of course, could see it a mile off;
she was shouting it from the rooftops. A child was seeing it, and
telling us how it was. Climate anxiety grew in society, and this can
be extremely harmful – for the individuals themselves, because we
know that anxiety and trauma impact on both physical and mental
health, and for the planet, when that anxiety becomes maladaptive,
when 'eco-paralysis' sets in, when we feel there's nothing we can do,
that there's no hope. We stop trying.

That's why I wanted to write this book, and I wanted it to be
driven by emotion (my emotions) as well as critical thinking. That's
why I started with the mud of Valencia and a story about those
working-class lads from Belfast (from my streets) terrorised and
traumatised by that dreadful war and then punished by so-called
therapists for their reaction when they returned home. I think it is a
truly shocking story. But for young people today faced with climate
change, and the inevitability of more and more extreme weather
events, and then called 'snowflakes' or neurotic when they express
their anxieties, one must conclude that this is equally shocking.

At least the First World War finished after four years, this new
crisis will not be so brief.

I have spent a lot of time in this book trying to locate climate
anxiety in context, or rather various contexts: a historical context
in the sections on shell shock and PTSD (new traumas connected to
changed and unusual conditions can be slow to be formally recognised); a clinical context (in terms of how climate anxiety sits with
respect to clinically-recognised disorders); a discursive context (in
terms of how people talk, gossip and socially construct climate anxiety and its sufferers in language); a social identity context (in terms
of climate deniers, their sense of belonging and their rhetoric); an
international business/financial context (in terms of the roots of climate denial); a personality context (in terms of individual difference

in how we process messages); a socio-political context (in terms of ongoing and competing global crises); a neuroscience context (in terms of how the brain works); and last but by no means least a personal context. Psychologists are after all people with the same worries, concerns and anxieties as everyone else. But, of course, I had to be critically aware throughout, sometimes playing the role of the detached academic, sometimes less so.

So how should we understand climate anxiety? As a psychologist, I needed to start with a fundamental issue, which is clearly a major dilemma for many researchers, clinicians and ordinary members of the public. Climate anxiety is not included in the latest *Diagnostic and Statistical Manual* (DSM-5-TR). Many people think that this is a very good thing. Britt Wray (2023) has written, 'many mental health professionals say it is important that it remains excluded. After all, the last thing we want is to pathologize this moral emotion, which stems from an accurate understanding of the severity of our planetary health crisis' (Wray 2023: 21). But this creates this essential dilemma, as van Valkengoed (2023) and others have noted:

> Being in a constant state of distress is neither desirable nor realistic, and finding ways to reduce climate anxiety should therefore be an important goal for scientists and practitioners alike. While climate anxiety is not a mental health problem, it is urgent and necessary that we start treating it like one.
>
> (van Valkengoed 2023: 387)

She pointed out that there was very limited research on psychological interventions to help sufferers cope better and no randomised control trials to date (as of 2023). Official recognition would surely change that. We need good science to allow us to help those who suffer from climate anxiety.

I argued right at the beginning of the book that climate anxiety can be a debilitating psychological condition. Linking a serious mental condition to accurate appraisal of a real cause would be a very powerful message about climate change and its effects. It might help some take climate change more seriously and that may be extremely beneficial in and of itself. It would certainly encourage research on how to alleviate it, and given that feelings of hopelessness are often a core aspect of climate anxiety, it could perhaps help sufferers deal more effectively with their 'eco-paralysis' and use some of the negative emotions associated with climate anxiety, like anger, to drive them to more effective climate action, to transform maladaptive anxiety into adaptive anxiety (Taylor 2020).

But then you're labelling a rational response to an irrational situation as a mental disorder, and that could be extremely damaging.

Given that climate anxiety isn't included in DSM-5, it's defined in a variety of ways and researchers use differing criteria to identify it. I spelt out the implications of differing dictionary definitions, as well as various operational definitions in the psychological literature. Climate anxiety is variously defined as 'unease or apprehension about current and future harm to the environment caused by human activity and climate change', 'a state of distress caused by concern about climate change', 'a condition in which someone feels frightened or very worried about climate change'. The American Psychological Association uses the term 'chronic fear' with its emphasis on persistence in the use of the word chronic. State of distress, on the other hand, doesn't emphasise persistence, 'states' by definition are temporary. The *Cambridge Dictionary* introduces 'very worried about climate change' compared with 'worried' as part of apprehension in the OED's definition. Fear and worry are different, worry is more cognitive (it is about thinking), fear is an emotion (like anxiety) with a physiological component (heart beating faster, palpitations, butterflies in the stomach). Some of the leading anxiety theorists view *worry* just as one part of anxiety, which potentially helps the sufferer deal with the anxiety, as we shall see. Fear and anxiety are both *felt*. In the book, I addressed the issue of feelings about climate change (and not just thoughts) at length because I think they are critical.

This variability does not help and not surprisingly incidence of climate anxiety does vary significantly from one study to another depending upon which definition is employed, but there is very clear evidence of major psychological distress about climate change (and its consequences) in a significant proportion of the population. Depending upon how the condition or state is defined we may think about different causal factors and developmental pathways.

I began by focusing on one particular contemporary manifestation of climate change, the extremely heavy rain in Valencia in Spain in November 2024, with ordinary citizens digging fellow citizens out of the mud. Shocking and alarming images. No doubt there will be many in that region who will suffer from the symptoms of PTSD in the years ahead, because of their alarming and traumatic experiences in the extreme weather brought on by climate change, and the fear and uncertainty about the future. I drew a parallel with what happened to soldiers in the First World War, describing in detail the experiences of soldiers trapped in the mud of the trenches (it wasn't just the mud that was the common element between the two situations), shelled night and day, where neither fight nor flight (our evolutionary primitive responses) was possible. Shell shock and climate anxiety share several characteristics including victim blaming, indeed victim denigration, lack of understanding of the condition, and often a complete lack of sympathy for the sufferers. Both conditions have been viewed as a sign of weakness or worse ('weak stock' in the case of shell shock; the 'snowflake' generation in the case of climate anxiety). I argued that we need to learn from past mistakes and take a much more sympathetic approach.

I am not saying that climate anxiety is necessarily felt as intensely as shell shock, or even that it will necessarily be similar to PTSD, but it is likely to be so for those who experience it directly (like those in Valencia in those terrible floods). In other words, climate anxiety may well include those who suffer from diagnosable PTSD at one extreme. But trauma resulting from climate change is always going to be different from the trauma of war. Those soldiers in the First World War returned home. The source of their trauma was in the past (but relived sometimes on a daily basis). With PTSD resulting from extreme weather and climate change, the source of the trauma is still there, in front of them, and may always be there. That's what makes it different.

The nervous systems of many soldiers in the First World War were shattered by trench warfare and when the soldiers broke down, as many did, they were blamed for their failure to cope with the terror of these novel and inhuman conditions (so graphically described by the poet Siegfried Sassoon). Mutism was the most common symptom of shell shock (but also flashbacks, tics, hypervigilance, etc.) and ordinary rank-and-file soldiers were 'treated' with electric shocks back in England to get them to make the noise 'ah', again and again. This is a form of aversion therapy, although I hate using that word 'therapy' for such barbarism. After completion of the therapy, the 'patients' still weren't allowed to talk about what they had been through or shake hands with their 'kindly therapist', who clearly regarded them as less than men, deserving no respect or understanding. Some officers suffering from shell shock, however, were treated by Rivers at Craiglockhart Hospital outside Edinburgh in a much more humane and meaningful way, where the therapist got them to talk about their experiences and tried to help them restructure their thoughts to make them more acceptable. Of course, we now know that shell shock is a form of post-traumatic stress disorder, and we know how that affects the brain and the nervous system. We are much more sympathetic to their condition and contemporary therapies have built on the pioneering work of Rivers and others. This type of narrative therapy may be of relevance to help people more effectively with climate anxiety.

But climate anxiety is even more controversial than shell shock because so many people simply don't believe in anthropogenic (man-made) climate change. I have spent a lot of time in the book explaining why this climate change denial has occurred, and this involved me discussing cognitive biases, the maintenance of mood state, personality, personality disorder, and downright lies, from information processing issues to big business post-Kyoto. But this also involved a consideration of how the brain works, and the fact that there are two main systems of cognition, System 1 and System 2, one automatic and unconscious working on the principle of associative learning, and one slower and conscious and working on more logical principles. Human beings do suffer from a number of cognitive biases and I show how these biases can influence what information we extract about climate change. Optimists unconsciously see one thing, pessimists see another. They remember different information from the same articles and build their representation of the climate change 'problem' (or non- problem) on the basis of these memories.

The fact that big business constructed an alternative discourse about climate change ('unproven, weak science, biased science') and therefore manipulated the media to present 'balanced' pieces about climate change (with climate change believers asked to debate with climate change deniers) was disgraceful. Climate change deniers have carried their fight all the way into the present and into the future, with Donald Trump now as the President. People with certain personality types (who need to maintain a positive mood state) will always be drawn to his arguments. 'The world is not going to end, we don't have to seek difficult and costly solutions. Let's frack, baby, frack!' That is a very 'positive' message. It just happens to be false, extremely irresponsible and exceptionally dangerous.

If we are to do anything to mitigate the effects of climate change, we all need to be involved and that's why climate anxiety is so important. Climate anxiety can be a major barrier to this, involving, as it does, learned helplessness ('we've tried in the past and it didn't work so we've given up'), and feelings of hopelessness and despair. Agency is critical. We need to feel empowered. Climate anxiety (if we properly understand it as a form of anxiety) is about thoughts and feelings and we need as individuals to deal with these and integrate them. Climate anxiety continues to grow (Wray 2022), and can be overwhelming and induce a form of psychological 'eco-paralysis', impacting on both sleep and daily activities (Verplanken et al. 2020) as a result of frequent rumination (Verplanken and Roy 2013).

Reducing climate anxiety, and helping people deal more effectively with their maladaptive negative emotions regarding climate change, is a pressing issue for us all.

I pointed out that climate anxiety is hard to pigeonhole unambiguously using current recognised clinical disorders. The American Psychological Association defines climate anxiety as 'a chronic fear of environmental doom'. It's chronic, in other words, constant and strong, not short-term and therefore different from normal anxiety. Rather it's more long-term like generalised anxiety disorder (GAD). But GAD is viewed as having internal causes (genes, experiences, abuse, etc.), that's why it's labelled as 'generalised', it generalises to a range of outside potential threats. But climate anxiety has an external cause, namely climate change, which does have multiple manifestations, including most recently for me the mud of Valencia, with images in my head, a few recurrent dreams of suffocating in that mud, and flashbulb memories of the faces of those shovelling the mud, and what they found buried there.

It's similar in some ways to a number of clinical anxiety disorders, including normal anxiety (but it is chronic not temporary), GAD (but it is external not internal), and PTSD, certainly for those directly experiencing it but perhaps even for those not directly experiencing it. Climate anxiety is also experienced by many people who do not suffer from existing mental health issues or particularly strong anxiety sensitivity, so-called trait anxiety (as Kurth and Pihkala (2022) point out).

It's a new type of anxiety shaped by our modern world and caused (one might add) by the selfish pursuits of status, wealth and power through high carbon economies and industries that have flourished ever since the industrial revolution, and aided and abetted by commercial advertising.

Climate anxiety manifests itself in a number of different ways and the label 'climate anxiety' is consequently applied to a very wide range of ecologically-oriented affective experiences (Kurth and Pihkala 2022). There are clearly very negative psychological effects of climate change, but a diverse range of effects are currently subsumed under a single label 'climate anxiety'. It can clearly be considered a rational response to an irrational world (that would be my position), and perhaps even a necessary condition to start doing something serious about climate change itself. If it is a real *anxiety* (body and mind) and not just rumination or worry, not just thinking, but an actual somatic response, a feeling, this may guide you to positive action. Indeed, Sangervo et al. (2022) found that in Finland, their measures of climate anxiety (including worry and stronger manifestations of anxiety including the effects of climate emotions on functionality and psychosomatic symptoms) were positively correlated with climate action, which they say shows the value of climate anxiety as an *adaptive* response.

However, one troubling feature of the research on climate anxiety is the significant variability in the incidence rates reported. Some people find this very confusing. It leads to major disagreement about how common it really is. Is it something nebulous and flaky, hard to measure, hard to detect? This is what climate deniers love to hear. I am sure Donald Trump rubs his hands together when he sees that estimates of climate anxiety range from 9% to 56%. But this is attributable to the operational definitions used, with significant variability in the definition (and breadth) of what constitutes climate anxiety.

Defining climate anxiety as 'a chronic fear of environmental doom' captures the idea of an anticipatory emotion about something with an uncertain outcome. This definition parallels the NHS definition of anxiety – it's an anticipatory emotion, with negative affect about the future – 'a feeling of worry, nervousness, or unease about something with an uncertain outcome'. It's an anxiety response to something in the future. Pihkala (2020) has made a similar point – 'Among general scholars of anxiety, a basic and shared view of anxiety is that it is future-oriented and related to a threat about which there is significant uncertainty' (Pihkala 2020: 1). Other researchers have used quite different definitions, and some have broadened the concept considerably.

Searle and Gow (2010) measured climate change *distress* with twelve items focusing on the various negative emotions associated with climate change. They distinguished two factors, the first factor encapsulating what they called climate change *anxiety*, in response to questions like 'Thinking about climate change now makes me feel – tense, anxious, worried, angry, concerned, stressed, sad, scared, depressed.' The second factor was labelled climate change *hopelessness* ('Thinking about climate change now makes me feel powerless, helpless, hopeless'). You can see that in the first factor climate anxiety per se is just one of a series of items; you can also see that this really is only about the negative *emotions* associated with thinking about climate change.

Reser et al. (2012) defined climate change distress as 'experienced apprehension, anxiety, sorrow, or loss due to the threat and projected consequences of climate change, for oneself, humanity, and/or the natural world'. They found that younger people and women showed higher levels of distress. This has the anticipatory element inherent in definitions of anxiety but includes other emotions, quite different from anxiety, like 'sorrow' and 'loss'.

In a systematic scoping review in 2021, Coffey et al. concluded that:

> Our review revealed a lack of clarity about the concept of eco-anxiety. For instance, there is a range of terms that overlap or are closely related between anxiety, dread, grief, worry, fear and despair … There is a call for further work on defining the terms with the need to raise discussion about the various aspects of eco-anxiety to diminish misunderstandings.
>
> (Coffey et al. 2021: 4)

And this is not just a semantic quibble. If different studies use completely different definitions of the condition, then when they measure the incidence of climate anxiety in the population, these estimates will vary considerably. And they do. This variability might suggest to some sceptics that this is a made-up, 'manufactured' disorder, a trendy gimmick, nothing less. Lack of consistency (and rigour) in the definition also means that it is hard to compare the relative incidence of climate anxiety in different demographics across studies.

Once defined consistently, we need to know how it connects to actual psychological distress. Clayton and Karazsia (2020) pointed out that little of the previous work had examined the relationship between climate anxiety and psychological well-being, which they aimed to rectify, adapting items from various clinical tests including rumination measures (to determine if people were thinking about climate change to an unhealthy extent) and a functional impairment measure (to see if climate change was interfering with people's ability to function). They also employed a measure of a general tendency to anxiety and depression, and measures of experience of climate change and behavioural engagement.

They found that depression and anxiety were strongly correlated with cognitive impairment (e.g. 'Thinking about climate change makes it difficult for me to concentrate') and functional impairment (e.g. 'My concerns about climate change undermine my ability to work to my potential'), but only moderately related to experience ('I have been directly affected by climate change') and not related at all to behavioural engagement ('I try to reduce my behaviours that contribute to climate change'). Importantly, the authors say that the negative emotions in response to climate change (feeling sad, scared, angry, etc.) can be distinguished from a more clinically significant anxious response. Negative emotions were associated with behavioural engagement, whereas climate change anxiety was not.

This study was clearly a very important step, but there are a few issues here. The US-based sample was relatively small (a few hundred) and perhaps not representative of the population as a whole. Furthermore, a general survey like this may not have reached a population genuinely suffering from any significant degree of climate anxiety. The scores on the anxiety subscales were low throughout, although 25% of the sample did report *some* degree of functional impairment (but that just means a response greater than 'sometimes'). We need more of a focus on specific groups, like those who might have sought professional or medical help for anxiety related to climate change, to understand the relationship between these different factors. The survey was online and the researchers had to screen out participants who took less than 90 seconds to complete the test (that is very quick indeed given the number of questions, all of which require some thought). How many participants took only slightly longer than 90 seconds, perhaps indicating a lack of serious-ness and commitment?

But there are other problems – some of the measures perhaps don't measure what they had been designed to assess. For example, 'I believe I can do something to help address the problem of climate change' is not necessarily a measure of 'behavioural engagement' (as suggested in the paper) but more likely a measure of self-efficacy (belief in one's ability to influence things). Similarly, 'I wish I behaved more sustainably' again doesn't necessarily measure behav-ioural engagement. You might be doing lots of sustainable things (i.e. have very high behavioural engagement) but wish that you were doing even more. Also, how can we assess whether the self-reports of actual behavioural engagement ('I recycle') were accurate? What does 'I recycle' actually mean? I'm sure we have all recycled something in our lives. How often do you have to do it, in order to answer yes to this particular item? I recycle some things. Does that count? Or do you have to be a habitual dedicated recycler? You can't leave it up to the participants to decide on the meaning of the question because if you do, you will get significant variability in interpretation.

Some of the measures of cognitive-emotional impairment were what look much more like coping mechanisms ('I write down my thoughts about climate change and analyse them'), some were not ('I find myself crying because of climate change'), but they are lumped together as measures of cognitive-emotional impairment (rather than say cognitive-emotional activity). The relationship between the neg-ative emotions and anxiety needs more work to tease out because the negative emotions are combined but some may be more instrumen-tal in the development of anxiety, indeed some feelings/emotions like 'worry' represent an important stage in the appraisal of threat and the intensity of the anxiety experienced, as some of the leading anxiety researchers like Aaron Beck have suggested.

Clearly, we need additional research based on the great start made in Clayton and Karazsia's paper – finer-grain analysis of emo-tions, more intensive questioning about behavioural engagement with evidence of this, as well as some modelling of how emotional responses and feelings of climate anxiety fit together in causal models, as has been done in models of anxiety and generalised anxiety disorder (rather than just factor analyses and a series of correlations, identifying associations, but nothing more).

Sometimes understanding climate anxiety means understanding the potential shortcomings in the published research and recognising that there's a lot more work to be done here, and that's a challenge for us all. Pihkala (2020) has gone so far as to suggest that 'climate change distress' might be a better term than 'climate anxiety' because it might be more accurate, and it is reminiscent of the research area of 'environmental stress' where the impacts of various environmental conditions on health have been researched for years. This new title, because of its semantic connotations and subtle links with the concept of 'environmental stress', does point to external and environmental factors. Climate anxiety, on the other hand, does seem to suggest something internal and maladaptive.

MOVING FORWARD WITH CLIMATE ANXIETY

Following on from van Valkengoed (2023), there are now a small number of articles being published recommending different sorts of psychotherapeutic interventions for climate anxiety. They look promising but they tend to be very small in scale. For example, Lindhe et al. (2023) used (internet-delivered) cognitive behavioural therapy (ICBT) to alleviate distress associated with climate change. They described this as a pilot study and it involved sixty participants, thirty in the experimental condition and thirty in the control condition. Measures of depressive symptoms, stress and quality of life were used as the primary outcome measures. The therapy consisted of informative texts and CBT exercises focusing on behavioural engagement, nature connectedness, acceptance, self-esteem and self-compassion, anxiety and constant worry, relaxation, cognitive restructuring, emotion regulation, perceived inadequacy, social relationships, sleep strategies, stress management and loneliness.

In other words, a very broad range of thoughts, feelings, strategies and actions. They found significant differences between the experimental and the control group on all three principal primary outcome measures and concluded that individually tailored ICBT can be an effective way to reduce psychological distress associated with climate change, but very importantly without reducing (self-reported) pro-environmental behaviour. This is a very promising study because ICBT can clearly be rolled out across different populations and is relatively economical in terms of therapist involvement.

Budziszewska and Jonsson (2022), on the other hand, analysed psychotherapeutic conversations with patients suffering from climate anxiety using a qualitative approach called Interpretative Phenomenological Analysis (IPA). One of their main findings was that therapists' knowledge about climate change and their ability to validate climate change emotions in their patients were perceived as very beneficial. But this was a small highly-selected sample of ten individuals.

Clearly, there is a lot of new research to be done in this area to develop and evaluate psychotherapeutic programmes (with appropriate randomised control procedures, the gold standard of psychological and medical research). But there are now several self-help guides, each with its own distinctive merits, including Wray (2022), Ray (2020), Kennedy-Woodard and Kennedy-Williams (2022), Verlie (2022) and Grose (2020), which some readers may well find useful.

I want to add to this by suggesting two small things which may help with climate anxiety on the basis of the work that I've been doing. Let me warn you that they may well sound trivial! Both are based on individual actions, the things we can control, things that we can do immediately. One involves listening, the second involves writing.

The first involves listening to our children, the children of our community, and understanding their fears and anxieties about the future under climate change. This can make climate change somatic, it will take the concept of climate change well beyond the cognitive domain, it will affect how you feel. Listening to children can be a sharp lesson for System 1, the system that guides our automatic choices. The second involves breaking out of the silence that haunts people, it involves disinhibition and the construction of narrative – writing about your own fears and anxieties about climate change. Dealing with your own trauma by externalising it, giving it meaning through language. Research has consistently shown that inhibiting thoughts and feelings requires a great deal of effort in the form of increased autonomic nervous system activity, which is inherently stressful and makes you more susceptible to stress-related diseases. Disinhibition is hugely beneficial for our health. But as well as disinhibition per se, forming a narrative about emotional trauma allows you to organise your complex and fractured emotional experiences and integrate your thoughts and feelings. The structure and meaning of the narrative can make the traumatic experience more 'manageable', facilitating a sense of resolution and resulting in less rumination (and fewer sleepless nights).

My first suggestion is based on the assumption that if we did more to tackle climate change itself then that would help with climate anxiety. There was a critical argument put forward in this book centring on the 'value-action' gap, the gap between what people say and what they do. We see it in others and some of us (unfortunately) can see it in ourselves. Most of us know that living sustainably is a good thing, and when asked we're more than happy to report our positive attitudes to sustainability and low carbon lifestyles and products. But there is another side to all of us, an unconscious side (that System 1 again), affecting what we notice, governing our quick decisions in the supermarket, and whether we jump on a bus or get into the car, governing the very stuff of life.

But can we make our implicit attitudes to low carbon products and lifestyles more positive, can we change how people feel about sustainability?

Laura McGuire and I carried out one study which was the first to test whether we can experimentally influence implicit attitudes to low carbon products, i.e. make these implicit attitudes more positive and therefore more likely to lead to more sustainable behaviour. Previous research had demonstrated that you can influence implicit attitudes to race (implicit attitudes of White people tend to be more positive to White people than to Black people: see Beattie 2013). The critical variables in the race modification research were *emotion* and *engagement* with the narratives used (Lai et al. 2014). All the successful interventions were those that encompassed both self-involvement and evaluative conditioning. Implicit attitudes to race changed only with interventions where participants were asked to imagine *themselves* as the victim of an assault in the story with a White attacker (high self-involvement), as opposed to merely imagining someone else as the victim of an assault, and, in addition, where Black people were associated with positivity and White people with negativity (a form of evaluative conditioning).

So, in our study we used highly emotional and engaging film clips from Al Gore's *An Inconvenient Truth* (see also Beattie 2011b). We discovered that we were able to influence implicit attitudes to low carbon products and this modification was associated with an increase in low carbon choices in an experimental task (Beattie and McGuire 2020).

These implicit attitudes may, however, change for one of two reasons – there may be a change in underlying associative structure (the underlying set and structure of associations), or there could be a change in the 'activation of pre-existing patterns', without necessarily any change in the underlying associative structures (Gawronski and Bodenhausen 2006). In other words, some sort of priming effect like in the 'banana-vomit' 'experiment' in Chapter 6. Given how brief the study was, it's much more likely to be just priming rather than change in underlying associative structure. Our view was that much longer-term *repetition* of these sorts of emotionally-laden images and narratives would be necessary to change the underlying associative structures in relation to low carbon. After all, the conceptual foundation of advertising is 'repetition, repetition, repetition'.

So we tried a more systematic approach to this issue. Laura developed and ran a series of educational programmes (between 2020 and 2023) for schoolchildren in twelve primary schools in areas of Merseyside in the North West of England, with schoolchildren aged 8–9 (n=378). A critical age for these children, when their associative structures (and understanding of consumer culture) relating products/lifestyles and sustainability are being established.

She ran two different climate change education programmes – a traditional knowledge-based teaching programme and a creative arts programme (with drama, drawing, posters, poetry writing, rap, etc. about climate change), plus a control programme that didn't involve discussion of climate change. At the beginning and end of the programmes, to assess their effectiveness, she measured feelings about climate change, explicit and implicit attitudes to carbon, and both reported behaviours and behavioural choices. She found that the more emotional and engaging creative arts-based programme was significantly more effective in changing the children's *feelings* about climate change and their sustainable *behaviours* (turning off the lights when not in the room, turning the heating down, etc.). Children were putting Post-It notes by the light switches at home reminding their parents to turn the light off. Some parents just didn't get it and occasionally there were some quite heated family 'discussions' about this.

But these children were great ambassadors for change in their own families and we started to wonder whether they could be the agents of change in a bigger way. And what was there to lose? Various 'top-down' campaigns about climate change have employed politicians, scientists and celebrities to deliver the message, but all with limited success. Many of these spokespersons were perceived to have a particular economic or ideological agenda (and some like Al Gore had been more or less discredited as role models). They seemed to be talking down to ordinary people. Could children from these working-class communities be more effective ambassadors than all these celebs?

EMPOWERMENT TOWARDS CLIMATE ACTION

With support from the Arts and Humanities Research Council in the UK, we made a short documentary film ('The Great Community Climate Change Experiment'), using a slightly older cohort of schoolchildren (11–12 years old). The film showed the creative arts education programme in action, and Laura's great teaching, and the process of empowerment and change in the children, as they talked about their changing feelings about climate change and explained what they were now doing to live more sustainably. They talked very openly about their fears and anxieties about their future under climate change, and what they were trying to do around the home to influence their parents (some of it was quite heartbreaking). In the film they display their art work and posters, and talk about what climate change means to them. They read their poetry. It is the poetry of children speaking directly. Childlike, naïve, fresh, real. People told us it was a very moving film. One senior member of our university was moved to tears.

We then put the film online (with financial support from the British Academy) and measured the effects of this film on adult audiences from the same working-class communities in the North West (mainly the area around Liverpool), assessing their attitudes to carbon, their feelings of empowerment (self-efficacy and response efficacy) versus learned helplessness, as well as their personal responsibility to act. We measured their perceived personal vulnerability and their emotional feelings towards climate change, as well as their beliefs and feelings of morality about climate change. And finally, we assessed their reported behaviours and behavioural intentions. There was a control group with a neutral film (a film about crocheting).

We got some striking results. We found that the experimental group of adults, watching the film of the children talking about their emotions and actions, showed a statistically significant (146.9%) increase in strong feelings of *empowerment* compared to the control group, and a significant (66.9%) decrease in feelings of learned helplessness. We analysed several different emotional responses to climate change following the film. The specific negative emotions analysed were sadness, fear, anger and disgust. There was a statistically significant 68.9% increase in negative emotions in the experimental group after watching the film compared with the control group. Interestingly, the biggest statistical effects were for *anger* and for *disgust*.

In other words, the adults were angrier and more disgusted by climate change having witnessed these children from their own community talking about their fears and anxieties about climate change, and what they were trying to do to help in their own small way. The adults felt more empowered and less helpless – having watched their children engage. We were delighted with these results and felt that this research could have implications for the design of future climate change campaigns with more of a *local* emphasis in the messaging, where the credibility of the message is enhanced by showing how change can happen within that community with children taking the lead. They after all are the future.

The conclusion is simple – listening to our children can affect how we think and *feel* about climate change. Let them tell you about their emotions, ask them to write some poetry to express it, or paint a picture. Encourage them to rap about how they feel about climate change, and what they've been trying to do to help with this. No constraints, no barriers, no defences. Soak it all in. It empowers us (that's the strange thing), and makes us more determined to do something. It is visceral. It makes us angry and disgusted and, as we have seen throughout this book, how we *feel* is critical to behaviour change and greener actions. Anger itself can be a great motivator when it is channelled appropriately. Then talk to your friends and neighbours about how to channel it appropriately through communities and joint action, and get together to put pressure on those above. A grassroots revolution from the bottom up!

There are grounds for being optimistic that small changes like this might make a difference. A study from the University of California Davis simulated 100,000 possible future policy and emission trajectories to identify those variables that could impact on climate change. Their conclusion was that '*public perceptions of climate change*, the future cost and effectiveness of climate mitigation and technologies, and how political institutions respond to public pressure are all important determinants of the degree to which the climate will change over the 21st century' (Moore et al. 2022, my italics; see also press release UC Davis, 16 February 2022). The lead author Frances

C. Moore from the Department of Environmental Science and Policy at UC Davis is quoted in the press release as saying:

> Small changes in some variables, like the responsiveness of the political system or the level of support for climate policy, can sometimes trigger a cascade of feedbacks that result in a tipping point and drastically change the emissions trajectory over the century.

But it is not just public *perceptions* that will matter here (no matter how thoughtful or considered), as we have seen, it is our intense *feelings* about climate change that will produce disruptive action, with change in ourselves, change in public support for climate policy, and change in the pressure (now more intensely felt) that we can exert on our political institutions. And these *feelings* can be changed. We now know that. Our children can change them; they can change us – when we listen. We *can* reach a tipping point beginning with these small changes, ripples turning into cascades of feedback, ultimately crashing onto our political institutions.

CLIMATE ANXIETY AND NARRATIVE

But we also have to look after ourselves throughout all of this; we need self-care. We know that different psychological factors influence the nature and level of climate anxiety experienced, including age, social disempowerment due to class or ethnicity, geographical location, as well as factors linked to emotionality and personality (Beattie 2018b). But several psychological *processes* are also highly relevant to how we experience climate anxiety, including whether we inhibit our thoughts and feelings about climate change, or whether there is a disconnect between our thoughts and emotions. Both can have major effects on our mental and physical health. This takes us right back to the beginning of the book and shell shock, where ordinary soldiers were not just discouraged from talking about their war-time trauma but actively prevented from doing so. Their silences spoke volumes. They suffered in silence and they died in silence. It wasn't just my grandfather who died young. Evidence from clinical psychology (Pennebaker 1995) and medicine (Booth et al. 1997) suggests that these processes (inhibition, disconnection between thoughts and emotions, rumination) arising from traumatic experiences can be positively influenced by encouraging people to write about their thoughts and feelings.

Pennebaker's paradigm to reduce emotional inhibition, and facilitate assimilation or the 'making of meaning' regarding trauma, is a simple one. He gets people 'to explore their deepest emotions and thoughts about traumatic emotional experiences' in writing over three consecutive days for twenty minutes each day (300–500 words a day). The research finds improvements in levels of depression, lowered blood pressure and even improved cancer survival rates (Pennebaker 1993). These highly significant findings have been supported by several recent meta-analyses (Rude and Haner 2018).

Pennebaker suggests that personal narratives about traumatic emotional experiences necessitate giving these memories a sequential, organised structure and allow for the recognition, labelling and expression of the emotions concerning what happened (Pennebaker 1997). These are then encoded and stored in a more coherent and simplified manner, reducing rumination. Pennebaker finds that those who benefit most use high frequencies of positive emotion words ('happy, joy, peaceful'), moderate frequencies of negative emotion words ('sad, hate, hurt, guilty') and increase their use of insight words ('realise, understand, thought, knew') and causal words ('because, why, reason') on successive days, demonstrating the assimilation process in action. They are working out why they are feeling and acting in particular ways (using insight words like 'understand') and how things are connected (using causal words like 'because'), but trying to be positive about the future. Research also tells us that a reduction in intrusive and avoidant thoughts through such writing can have beneficial effects on adjustment by freeing up the cognitive resource of working memory, often significantly impacted by intrusive thoughts (Park and Blumberg 2002). Developing feelings of self-efficacy and empowerment in these personal narratives may also be critical to enhanced well-being (Beattie and McGuire 2018).

So, here is another simple tip – write about climate change and your fears and anxieties, your 'deepest emotions and thoughts', over three consecutive days for twenty minutes each day. This is a process of structuring, organising and assimilating thoughts and emotions. Do this every so often when in despair. Such a simple process but one that seems to work for many people.

Two simple steps and very easy to follow – listening to our children talking openly about climate change, and writing about our emotional traumas and anxieties about climate change. These can be (surprisingly) empowering actions and may allow us to cope more effectively with what we ourselves are currently experiencing. One is focusing on our feelings and actions and linking these two things, the other is focusing on our thoughts and emotions. Both are about empowerment, integration and appropriate action.

Perhaps I should finish by saying that writing this book in a more personal way has perhaps helped me deal with my own conflicted emotions (Beattie 2021). I wrote both about historical trauma, and my grandfather's generation and the silence that greeted these traumatised heroes on their return, and contemporary trauma, with climate change unfolding right in front of us, demonstrating its full awful power in the mud of Valencia. The heroic stories and the cultural memories of the Ulster Division in the trenches of the Somme (told and retold in my community growing up and to the present) and what I saw and learned about the psychological aftermath (long before I became a psychologist), with those old men that I met as a child, sitting alone and silent, stammering and blinking, in those damp grey streets of Belfast full of dilapidated mill houses. Emotional images that were hard to integrate, hard to reconcile, harder to understand. And without language and narrative, impossible. Writing about the mud of Valencia helped me articulate the feelings and thoughts of this (distant) observer about climate change, but not so distant that I couldn't see the trauma and appreciate their pre-traumatic stress reactions about the future trajectory of climate change.

Writing in this personal way was a step for me towards finding a voice, a way of articulating and sharing thoughts, of reducing emotional inhibition and facilitating assimilation or the 'making of meaning', of confessing my own anxieties and fears, and sometimes my own (embarrassing) distrust of people (and myself!). I am nothing if not fair. The value-action gap does, after all, cast a long and ominous shadow for those of us worried about the consequences of climate change.

We can each of us try to engage in this emotional disinhibition in our own way and I would encourage you all to try. It is, after all, a way of expressing our own *psychological truth* about the emotional cost of climate change and, at the same time, articulating the real *physical truth* of climate change, before it's too late. We know that many things have to change, but change needs to start with us. Only then can a tipping point be reached with regard to climate change and only then could, and should, climate anxiety begin to dissipate.

FURTHER READING

Beattie, G. and McGuire, L. (2020). The modifiability of implicit attitudes to carbon footprint and its implications for carbon choice. *Environment and Behavior* 5: 467–494.

Coffey, Y. et al. (2021). Understanding eco-anxiety: A systematic scoping review of current literature and identified knowledge gaps. *The Journal of Climate Change and Health* 3: 100047.

Grose, A. (2020). *A Guide to Eco-Anxiety: How to Protect the Planet and Your Mental Health.* London: Watkins.

Moore, F. C. et al. (2022). Determinants of emissions pathways in the coupled climate-social system. *Nature* 603: 103–111.

Pennebaker, J. W. (1995). *Emotion, Disclosure and Health.* Washington, DC: American Psychological Association.

Pihkala, P. (2020). Anxiety and the ecological crisis: An analysis of eco-anxiety and climate anxiety. *Sustainability* 12: 7836.

Ray, S. J. (2020). *A Field Guide to Climate Anxiety.* Oakland: University of California Press.

Sangervo, J. et al. (2022). Climate anxiety: Conceptual considerations and connections with climate hope and action. *Global Environmental Change* 76: 102569.

Verlie, B. (2022). *Learning to Live with Climate Change: From Anxiety to Transformation.* London: Routledge.

SUMMARY: THINKING OF THE FUTURE

- We need significant changes to mitigate the effects of climate change and we all need to be involved.
- Climate anxiety can be a major barrier to this, involving, as it does, learned helplessness ('we've tried in the past and it didn't work so we've given up'), and feelings of hopelessness and despair.
- Agency is critical.
- We need to feel strong and empowered.
- Climate anxiety (as a form of anxiety) is about feelings and thoughts and we need as individuals to deal with these and integrate them.
- Climate anxiety is growing.
- It can be overwhelming and induce a form of psychological 'eco-paralysis', impacting on both sleep and daily activities as a result of frequent rumination.
- Reducing climate anxiety, and helping people deal more effectively with their negative emotions regarding climate change, is a pressing issue for us all.
- Climate anxiety is experienced by many people who do not suffer from existing mental health issues or particularly strong anxiety sensitivity (so-called trait anxiety).
- Climate anxiety is hard to diagnose using current recognised clinical disorders.

- It's similar in many ways to some clinical anxiety disorders, including normal anxiety (but it's chronic not temporary), GAD (but it's external not internal), and PTSD, even for those not directly experiencing it (but often less intense).
- It's a new type of anxiety shaped by our modern world and caused by the selfish pursuits of status, wealth and power through high carbon economies and industries that have flourished ever since the industrial revolution.
- Operational definitions of climate anxiety vary widely and therefore so do reported incidence levels.
- We may need to consider an alternative label like 'climate change distress'.
- I outline two steps we can all take – one to help empower us with regard to climate change mitigation, the other to help us deal with our own internal turmoil.
- The first involves listening to our children talking openly about climate change to influence how we feel about climate change.
- Our feelings are crucial to our actions.
- And these feelings can be changed. We now know that.
- Our children can change us.
- The second involves us writing about our own thoughts, feelings and fears regarding climate change.
- This is a process of structuring, organising and assimilating thoughts and emotions with demonstrable benefits for our physical and mental health.
- Such a simple process but one that seems to work for many people.
- Like all new approaches, both have been around for some time (forever?), it's just that now we have some evidence of their efficacy.
- We know that many things have to change, but it needs to start with us.
- Only then can a tipping point be reached with regard to climate change and only then could, and should, the problems of climate anxiety begin to be remedied.

GLOSSARY

Affect: 'a feeling state either with or without consciousness ... Affective responses occur rapidly and automatically' (Slovic 2002).

Anxiety: a feeling of worry, nervousness or unease about something with an uncertain outcome.

Attitude: 'A mental and neural state of readiness to act' (Allport 1935).

Attributional style: particular patterns of thinking that we use to reason about why things have occurred in everyday life.

Big lie: a gross distortion or misrepresentation of the truth used as a propaganda technique.

Carbon labelling: information about GHG emissions in CO_2 equivalents that can be ascribed to goods and services. Consumers are informed of the environmental impact of products, for example, through a labelling scheme, enabling them to reduce the CO_2 emissions of their household by making simple and relatively small changes to their lifestyle.

Climate anxiety: 'a state of distress caused by concern about climate change' (*Collins Dictionary*).

Dispositional optimism: a personality characteristic that predisposes an individual to expect that good things, rather than bad things, will happen to them in life.

Dispositional pessimism: a personality characteristic that predisposes an individual to expect negative outcomes in the future.

Dual-process models of the brain: the theory that the brain has two main modes of operation – one being slow, rational and analytic and open to conscious introspection; the other process being much faster and automatic, more 'experiential', and derived from our everyday world of associations, actions and feelings.

Eco-anxiety: 'a chronic fear of environmental doom … The chronic fear of environmental cataclysm that comes from observing the seemingly irrevocable impact of climate change and the associated concern for one's future and that of future generations' (American Psychological Association 2021).

Eye-tracking: a procedure to measure the exact points of fixation of the eyes when viewing images, for example, images of products with carbon labels.

Functional Magnetic Resonance Imaging (FMRI): an experimental technique to measure the small changes in blood flow that occur with brain activity. It can help identify which parts of the brain are involved in various cognitive tasks.

Hot spot analysis in eye-tracking: a representation of the pattern of fixations in an eye-tracking study; different shades indicate intensity of fixation, i.e. the exact spots that the eyes focus on.

Hypervigilance: extreme or excessive vigilance. A symptom of PTSD.

Implicit attitude: 'Actions or judgments that are under the control of automatically activated evaluation, without the performer's awareness of that causation' (Greenwald et al. 1998).

Mental disorder: 'a clinically significant disturbance in an individual's cognition, emotional regulation or behaviour' (World Health Organization 2017).

Optimism bias: the tendency of people to be over-optimistic about their own future outcomes.

Post-Traumatic Stress Disorder (PTSD): an enduring anxiety caused by a traumatic and horrific event characterised by re-experiencing of the trauma through flashbacks (vivid intense memories that don't fade with time), nightmares, repetitive and distressing images or sensations, physical sensations such as sweating or trembling, emotional numbing, hyperarousal and hypervigilance (feeling 'on edge' and being on guard all the time), irritability, angry outbursts, insomnia and difficult concentrating.

Pre-Traumatic Stress Reactions: 'intrusive images and dreams about negative future events, accompanied by attempts at avoidance and increased levels of arousal ... disturbing future-oriented cognitions and imaginations as measured in terms of a direct temporal reversal of the conceptualizations of past-directed cognitions in the PTSD diagnosis' (Berntsen and Rubin 2015).

Stigma: a mark of disgrace associated with a particular circumstance, quality or person.

Subliminal: influence without conscious awareness.

Theory of Mind: the ability to represent the mental states of others to explain and predict their behaviour.

REFERENCES

Aaker, D. A. and Biel, A. L. (2013). Brand equity and advertising: An overview. *Brand Equity & Advertising*: 1–8.

Act on CO2 (2007). About Act on CO2. Retrieved from http://webarchive. nationalarchives.gov.uk/20101007164856/http://actonco2.direct.gov. uk/home/about-us.html. Accessed 23 August 2018.

Ajzen, A. and Cote., N. G. (2008). Attitudes and the prediction of behaviour. In W. D. Crano (Eds.), *Attitudes and Attitude Change*. New York: Taylor and Francis.

Albrecht, G. (2005). 'Solastalgia'. A new concept in health and identity. *PAN: Philosophy Activism Nature* 3: 41–55.

Allport, G. W. (1935). Attitudes. In C. Murchison (Eds.), *Handbook of Social Psychology*. Massachusetts: Clark University Press.

American Psychiatric Association (1980). *Diagnostic and Statistical Manual of Mental Disorders, Third Edition*. Washington, DC: APA.

American Psychiatric Association (2000). *Diagnostic and Statistical Manual of Mental Disorders, Fourth Edition*. Washington, DC: APA.

American Psychiatric Association (2013). *Diagnostic and Statistical Manual of Mental Disorders, Fifth Edition*. Washington, DC: APA.

American Psychiatric Association (2024). *Diagnostic and Statistical Manual of Mental Disorders, Fifth Edition, DSM-5-TR*. Washington, DC: APA.

American Psychological Association (2021). Addressing climate change concerns in practice. https://www.apa.org/monitor/2021/03/ce-climate-change. Accessed 19 November 2024.

Arceneaux, K. and Truex, R. (2023). Donald Trump and the lie. *Perspectives on Politics* 21: 863–879.

Arendt, H. (1951/2017). *The Origins of Totalitarianism*. London: Penguin.

Armitage, S. and Crawford, R. (Eds.) (1998). *The Penguin Book of Poetry from Britain and Ireland since 1945*. London: Viking.

Armstrong-Jones, R. (1917). The psychopathy of the barbed wire. *Nature* 100: 1–3.

Arnett, J. J. (2000). Optimistic bias in adolescent and adult smokers and non-smokers. *Addictive Behavior* 25: 625–632.

Balaskas, S. (2024). HEXACO traits, emotions and social media in shaping climate action and sustainable consumption: The mediating role of climate change worry. *Psychology International* 6: 937–976.

Barker, P. (1992). *Regeneration*. London: Penguin.

Baudon, P. and Jachens, L. (2021). A scoping review of interventions for the treatment of eco-anxiety. *International Journal of Environmental Research and Public Health* 18: 9636.

Beattie, G. (1992). *We Are the People: Journeys Through the Heart of Protestant Ulster*. London: Heinemann.

Beattie, G. (2004). *Protestant Boy*. London: Granta.

Beattie, G. (2010). *Why Aren't We Saving the Planet? A Psychologist's Perspective*. London: Routledge.

Beattie, G. (2011a). *Get the Edge: How Simple Changes Will Transform Your Life*. London: Orion.

Beattie, G. (2011b). Making an action film. *Nature Climate Change* 1: 371–373.

Beattie, G. (2012). Psychological effectiveness of carbon labelling. *Nature Climate Change* 2: 214–217.

Beattie, G. (2013). *Our Racist Heart? An Exploration of Unconscious Prejudice in Everyday Life*. London: Routledge.

Beattie, G. (2016). *Rethinking Body Language: How Hand Movements Reveal Hidden Thoughts*. London: Routledge.

Beattie, G. (2018a). *The Conflicted Mind: And Why Psychology Has Failed to Deal with It*. London: Routledge.

Beattie, G. (2018b). Optimism bias and climate change. *The British Academy Review* 33: 12–15.

Beattie, G. (2019). *Trophy Hunting: A Psychological Perspective*. London: Routledge.

Beattie, G. (2021). *Selfless: A Psychologist's Journey through Identity and Social Class*. London: Routledge.

Beattie, G. (2023). Doubt: A Psychological Exploration. London: Routledge.

Beattie, G. (2024a). Lies, Lying and Liars: A Psychological Analysis. London: Routledge.

Beattie, G. (2024b). Why does Donald Trump tell such blatant lies? The Conversation, 21 October.

Beattie, G. and Doherty, K. (1995). 'I saw what really happened'. Journal of Language and Social Psychology 14: 408–433.

Beattie, G. and Ellis, A. (2017). The Psychology of Language and Communication. London: Routledge.

Beattie, G., Sale, L., and McGuire, L. (2011). An inconvenient truth? Can a film really affect psychological mood and our explicit attitudes towards climate change? Semiotica 187: 105–112.

Beattie, G. and McGuire, L. (2014) The psychology of sustainable consumption. In A. Ulph and D. Southerton (Eds.), Sustainable Consumption: Multi-Disciplinary Perspectives. Oxford: Oxford University Press.

Beattie, G. and McGuire, L. (2015). Harnessing the unconscious mind of the consumer: How implicit attitudes predict pre-conscious visual attention to carbon footprint information on products. Semiotica 204: 253–290.

Beattie, G. and McGuire, L. (2016). Consumption and climate change: Why we say one thing but do another in the face of our greatest threat. Semiotica 213: 493–538.

Beattie, G. and McGuire, L. (2018). The Psychology of Climate Change. London: Routledge.

Beattie, G. and McGuire, L. (2020). The modifiability of implicit attitudes to carbon footprint and its implications for carbon choice. Environment and Behavior 5: 467–494.

Beattie, G. and Sale, L. (2009). Explicit and implicit attitudes to low and high carbon footprint products. International Journal of Environmental, Cultural, Economic, and Social Sustainability 5: 191–206.

Beattie, G. and Sale, L. (2011). Shopping to save the planet? Implicit rather than explicit attitudes predict low carbon footprint consumer choice. International Journal of Environmental, Cultural, Economic, and Social Sustainability 7: 211–232.

Beattie, G., Marselle, M., McGuire, L. and Litchfield, D. (2017). Staying over-optimistic about the future: Uncovering attentional biases to climate change messages. Semiotica 218: 22–64.

Beattie, G., McGuire, L. and Sale, L. (2010). Do we actually look at the carbon footprint of a product in the initial few seconds? An experimental analysis of unconscious eye movements. *International Journal of Environmental, Cultural, Economic and Social Sustainability* 6: 47–65.

Bechara, A., Damasio, H., Tranel, D., and Damasio, A. R. (1997). Deciding advantageously before knowing the advantageous strategy. *Science* 275: 1293–1295.

Beck, A. T. and Clark, D. A. (1997). An information processing model of anxiety: Automatic and strategic processes. *Behaviour Research and Therapy* 35: 49–58.

Berntsen, D. and Rubin, D. C. (2015). Pre-traumatic stress reactions in soldiers deployed to Afghanistan. *Clinical Psychological Science* 3: 663–674.

Berry, T., Crossley, D. and Jewell, J. (2008). Check-out carbon: The role of carbon labelling in delivering a low-carbon shopping basket. London: Forum for the Future.

Bettman, J. R., Luce, M. F. and Payne, J. W. (1998). Constructive consumer choice processes. *Journal of Consumer Research* 25: 187–217.

Blair, R. (1999). Responsiveness to distress cue in the child with psychopathic tendencies. *Personality and Individual Differences* 7: 135–145.

Blair, R. (2005). Responding to the emotions of others: Dissociating forms of empathy through the study of typical and psychiatric populations. *Consciousness and Cognition* 14: 698–718.

Blair, R. et al. (2005). Deafness to fear in boys with psychopathic tendencies. *Journal of Child Psychology and Psychiatry* 46: 327–336.

Bohm, G. (2003). Emotional reactions to environmental risks: Consequentialist versus ethical evaluation. *Journal of Environmental Psychology* 23: 199–212.

Bohner, G. and Dickel, N. (2011). Attitude and attitude change. *Annual Review of Psychology* 62: 391–417.

Booth, R. J., Petrie, K. J. and Pennebaker, J. W. (1997). Changes in circulating lymphocyte numbers following emotional disclosure: Evidence of buffering? *Stress Medicine* 13: 23–29.

Borkovec, T. D. et al. (1983). Preliminary exploration of worry: Some characteristics and processes. *Behaviour Research and Therapy* 21: 9–16.

Budziszewska, M. and Jonsson, S. E. (2022). Talking about climate change and eco-anxiety in psychotherapy: A qualitative analysis of patients' experiences. *Psychotherapy* 59: 606–615.

Campbell, J. et al. (2009). A behavioral genetic study of the Dark Triad of personality and moral development. *Twin Research and Human Genetics* 12: 132–136.

Chaiken, S. and Trope, Y. (Eds.) (1999). *Dual-process Theories in Social Psychology*.
London: Guilford Press.

Cianconi, P., Betrò, S. and Janiri, L. (2020). The impact of climate change on
mental health: A systematic descriptive review. *Frontiers in Psychiatry* 11:
490206.

Clayton, S. (2020). Climate anxiety: Psychological responses to climate change.
Journal of Anxiety Disorders 74: 102263.

Clayton, S. and Karazsia, B. T. (2020). Development and validation of a measure
of climate change anxiety. *Journal of Environmental Psychology* 69: 101434.

Coffey, Y. et al. (2021). Understanding eco-anxiety: A systematic scoping review
of current literature and identified knowledge gaps. *The Journal of Climate
Change and Health* 3: 100047.

Cohen, D., Beattie, G. and Shovelton, H. (2010). Nonverbal indicators of decep-
tion: How iconic gestures reveal thoughts that cannot be suppressed.
Semiotica 182: 133–174.

Crysel, L. C., Crosier, B. S., and Webster, G. D. (2013). The Dark Triad and risk
behavior. *Personality and Individual Differences* 54: 35–40.

Damasio, A. R. (1994). *Descartes' Error: Emotion, Reason and the Human Brain*. New York:
Putnam.

Danesi, M. (2020). *The Art of the Lie*. Lanham: Prometheus Books.

Darwin, C. (1877). A biographical sketch of an infant. *Mind* 7: 285–294.

DEFRA (2008). A framework for pro-environmental behaviours. Retrieved from
https://assets.publishing.service.gov.uk/government/uploads/system/
uploads/attachment_data/file/69277/pb13574-behaviours-
report-080110.pdf. Accessed 11 July 2023.

DEFRA (2016). Policy paper 2010 to 2015 Greenhouse gas emissions.
Retrieved from https://www.gov.uk/government/publications/2010-
to-2015-government-policy-greenhouse-gas-emissions/2010-
to-2015-government-policy-greenhouse-gas-emissions. Accessed 11 July
2023.

Department for Business, Energy and Industry Strategy (2017). Energy consumption in the UK. Retrieved from https://assets.publishing.service.gov.uk/
government/uploads/system/uploads/attachment_data/file/631146/
UK_Energy_in_Brief_2017.pdf. Accessed 11 July 2023.

DePaulo, B. et al. (1996). Lying in everyday life. *Journal of Personality and Social Psychology* 70: 979–995.

Downing, P. and Ballantyne, J. (2007). Tipping point or turning point. Social marketing and climate change. Ipsos MORI. Retrieved from https://www.ipsos.com/ipsos-mori/en-uk/tipping-point-or-turning-point-social-marketing-climate-change

Dreyfus, H. L. (1972). *What Computers Can't Do: The Limits of A.I.* New York: Harper and Row.

Eagly, A. H. and Chaiken, S. (2007). The advantages of an inclusive definition of attitude. *Social Cognition* 25: 582–602.

Ehrenreich, B. (2010). *Smile or Die: How Positive Thinking Fooled America and the World.* London: Granta.

Eisenman, D. et al. (2015). An ecosystems and vulnerable populations perspective on solastalgia and psychological distress after a wildfire. *EcoHealth* 12: 602–610.

Eisma, M. C. (2023). Prolonged Grief Disorder in ICD-11 and DSM-5-TR: Challenges and controversies. *The Australian and New Zealand Journal of Psychiatry* 57: 944–951.

Ekman, P. (2001). *Telling Lies: Clues to Deceit in the Marketplace, Politics and Marriage.* New York: Norton.

Fazio, R. (1990). Multiple processes by which attitudes guide behavior: The MODE model as an integrative framework. *Advances in Experimental Social Psychology* 23: 75–109.

Fourth National Climate Assessment (2018). US Global Change Research Program, Washington, DC.

Freud, S. (1949). *An Outline of Psychoanalysis.* New York: Norton.

Friese, M., Wänke, M. and Plessner, H. (2006). Implicit consumer preferences and their influence on product choice. *Psychology and Marketing* 23: 727–740.

Furnham, A., Richards, S. C. and Paulhus, D. L. (2013). The Dark Triad of personality: A 10-year review. *Social and Personality Psychology Compass* 7: 199–216.

Gadema, Z. and Oglethorpe, D. (2011). The use and usefulness of carbon labelling food: A policy perspective from a survey of UK supermarket shoppers. *Food Policy* 36: 815–822.

Gawronski, B. and Bodenhausen, G. V. (2006). Associative and propositional processes in evaluation: An integrative review of implicit and explicit attitude change. *Psychological Bulletin* 132: 692–731.

Gibson, K. E. et al. (2020). The mental health impacts of climate change: Findings from a Pacific Island atoll nation. *Journal of Anxiety Disorders* 73: 102237.

Gifford, R. (2011). The dragons of inaction: Psychological barriers that limit climate change mitigation and adaptation. *American Psychologist* 66: 290–302.

Gifford, R. et al. (2009). Temporal pessimism and spatial optimism in environmental assessments: An 18-nation survey. *Journal of Environmental Psychology* 29: 1–12.

Global Risk Report (2016). 11th Edition. www3.weforum.org/docs/Media/TheGlobalRisksReport2016.pdf. Accessed 10 July 2016.

Gounaridis, D. and Newell, J. P. (2024). The social anatomy of climate change denial in the United States. *Scientific Reports* 14: 2097.

Greenwald, A. G. and Banaji, M. R. (1995). Implicit social cognition: Attitudes, self-esteem, and stereotypes. *Psychological Review* 102: 4–27.

Greenwald, A. G. and Nosek, B. A. (2008). Attitudinal dissociation: What does it mean? In *Attitudes* (pp. 85–102). New York: Psychology Press.

Greenwald, A. G., McGhee, D. E. and Schwartz, J. L. (1998). Measuring individual differences in implicit cognition: The Implicit Association Test. *Journal of Personality and Social Psychology* 74: 1464–1480.

Greenwald, A. G. et al. (2002). A unified theory of implicit attitudes, stereotypes, self-esteem, and self-concept. *Psychological Review* 109: 3–25.

Greenwald, A. G. et al. (2009). Understanding and using the Implicit Association Test: III. Meta-analysis of predictive validity. *Journal of Personality and Social Psychology* 97: 17–41.

Gregersen, T. et al. (2024). How the public understands and reacts to the term 'climate anxiety'. *Journal of Environmental Psychology* 96: 102340.

Grose, A. (2020). *A Guide to Eco-anxiety: How to Protect the Planet and Your Mental Health.* London: Watkins.

Haidt, J. (2001). The emotional dog and its rational tail: A social intuitionist approach to moral judgment. *Psychological Review* 108: 814–834.

Hare, R. D. (1991). *The Hare Psychopathy Checklist Revised.* Ontario: Multi-health Systems.

Hartshorne, H. and May, M. (1928). *Studies in the Nature of Character, Studies in Deceit.* Columbia: Columbia University Press.

Hickman, C. et al. (2021). Climate anxiety in children and young people and their beliefs about government responses to climate change: A global survey. *The Lancet Planetary Health* 5: e863–e873.

Hine, D. (2023). *At Work in the Ruins*. New York: Chelsea Green Publishing.

Hitler, A. (1925/2022). *Mein Kampf*. Mumbai: Jaico Publishing House

Hoffman, A. J. (2015). *How Culture Shapes the Climate Change Debate*. Stanford: Stanford University Press.

Innocenti, M. et al. (2023). How can climate change anxiety induce both pro-environmental behaviours and eco-paralysis? The mediating role of general self-efficacy. *International Journal of Environmental Research and Public Health* 20: 1–10.

IPCC (1995). Climate Change 1995: The Science of Climate Change. https://efaidnbmnnnibpcajpcglclefindmkaj/https://www.ipcc.ch/site/assets/uploads/2018/02/ipcc_sar_wg_I_full_report.pdf. Accessed 20 November 2024.

IPCC (2001). Climate change 2001. https://www.ipcc.ch/site/assets/uploads/2018/03/WGII_TAR_full_report-2.pdf. Accessed 20 November 2024.

IPCC (2007). Climate change 2007: The Physical Science Basis. https://www.ipcc.ch/report/ar4/wg1/. Accessed 20 November 2024.

IPCC (2013). Climate Change 2013: The Physical Science Basis. https://www.ipcc.ch/report/ar5/wg1/. Accessed 20 November 2024.

IPCC (2014). AR5 Synthesis Report: Climate Change 2014. https://www.ipcc.ch/report/ar5/syr/. Accessed 20 November 2024.

IPCC (2023). Synthesis Report. https://www.ipcc.ch/report/sixth-assessment-report-cycle/. Accessed 20 November 2024.

Isaacowitz, D. M. (2006). Motivated gaze. The view from the gazer. *Current Directions in Psychological Science* 15: 68–72.

Jonason, P. K. and Webster, G. D. (2012). A protean approach to social influence: Dark Triad personalities and social influence tactics. *Personality and Individual Differences* 52: 521–526.

Kahneman, D. (2011). *Thinking Fast and Slow*. London: Penguin.

Kaplan, E. A. (2020). Is climate-related pre-traumatic stress syndrome a real condition? *American Imago* 77: 81–104.

Kennedy-Woodard, M. and Kennedy-Williams, P. (2022). *Turn the Tide on Climate Anxiety: Sustainable Action for Your Mental Health and the Planet*. London: Jessica Kingsley Publishers.

Kollmuss, A. and Agyeman, J. (2002). Mind the gap: Why do people act environmentally and what are the barriers to pro-environmental behaviour. *Environmental Education Research* 8: 239–260.

Kurth, C. and Pihkala, P. (2022). Eco-anxiety: What it is and why it matters. *Frontiers in Psychology* 13: 981814.

Lai, C. K. et al. (2014). Reducing implicit racial preferences: I. A comparative investigation of 17 interventions. *Journal of Experimental Psychology: General* 143: 1765–1785.

Lau, S. et al. (2024). Emotional responses and psychological health among young people. *Lancet Planetary Health* 8: e365–e377.

Lee, T. M. et al. (2015). Predictors of public climate change awareness and risk perception around the globe. *Nature Climate Change* 5: 1014–1020.

Lee, V. and Beattie, G. (1998). The rhetorical organization of verbal and nonverbal behavior in emotion talk. *Semiotica* 120: 39–92.

Lee, V. and Beattie, G. (2000). Why talking about negative emotional experiences is good for your health: A micro analytic perspective. *Semiotica* 130: 1–81.

Leed, E. J. (1979). *No Man's Land: Combat and Identity in World War 1*. Cambridge: Cambridge University Press.

Leiserowitz, A. (2005). American risk perception: Is climate change dangerous? *Risk Analysis* 25: 1433–1442.

Leiserowitz, A. (2007). Communicating the risks of global warming. In S. C. Moser and L. Dilling (Eds.), *Creating a Climate for Change*. Cambridge: Cambridge University Press.

Lindhe, N. et al. (2023). Tailored internet delivered cognitive behavioral therapy for individuals experiencing psychological distress associated with climate change. *Research and Therapy* 171: 104438.

Loewenstein et al. (2001). Risk as feelings. *Psychological Bulletin* 127: 267–286.

Lorenzoni, I. and Pidgeon, N. (2006). Public views on climate change: European and USA perspectives. *Climatic Change* 77: 73–95.

Luo, J. and Isaacowitz, D. M. (2007). How optimists face skin cancer information: Risk assessment, attention, memory, and behavior. *Psychology and Health* 22: 963–984.

MacDonagh, K. (1917). *The Irish on the Somme*. London: Hodder and Stoughton.

Machiavelli, N. (1513/1977). *The Prince*. New York: W.W. Norton.

Marshall, G. (2015). *Don't Even Think About It: Why Our Brains Are Wired to Ignore Climate Change*. London: Bloomsbury.

Matthews, A. and MacLeod, C. (2002). Induced processing biases have causal effects on anxiety. *Cognition and Emotion* 16: 331–354.

McCright, A. M. and Dunlap, R. E. (2011). The politicization of climate change and polarization in the American public's views of global warming, 2001–2010. *The Sociological Quarterly* 52: 155–194.

Meere, M. and Egan, V. (2017). Everyday sadism, the Dark Triad, personality and disgust sensitivity. *Personality and Individual Differences* 112: 157–161.

Moore, F. C. et al. (2022). Determinants of emissions pathways in the coupled climate-social system. *Nature* 603: 103–111.

Morganstein, J. C. and Ursano, R. J. (2020). Ecological disasters and mental health: Causes, consequences, and interventions. *Frontiers in Psychiatry* 11: 1–15.

Moser, A. K. (2015). Thinking green, buying green? Drivers of pro-environmental purchasing behavior. *Journal of Consumer Marketing* 32: 167–175.

Mosing, M. A. et al. (2009). Genetic and environmental influences on optimism and its relationship to mental and self-rated health: A study of aging twins. *Behavior Genetics* 39: 597–604.

Muller, G. E. and Pilzecker, A. (1900). Experimentelle beitrage zur lehre vom gedachtniss. *Zeitschrift fur Psychologie* 1: 1–300.

Ogunbode, C. et al. (2022). Climate anxiety, well-being and pro-environmental action. *Journal of Environmental Psychology* 84: 101887.

Ojala, M. (2015). Climate change skepticism among adolescents. *Journal of Youth Studies* 18: 1135–1153.

Orr, P. (1987). *The Road to the Somme: Men of the Ulster Division Tell Their Story*. Belfast: Blackstaff Press.

Pahl, S. et al. (2005). Comparative optimism for environmental risks. *Journal of Environmental Psychology* 25: 1–11.

Panzone, L. et al. (2016). Socio-demographics, implicit attitudes, explicit attitudes, and sustainable consumption in supermarket shopping. *Journal of Economic Psychology* 55: 77–95.

Park, C. L. and Blumberg, C. J. (2002). Disclosing trauma through writing: Testing the meaning-making hypothesis. *Cognitive Therapy and Research* 26: 597–616.

Pear, T. H. (1918). The war and psychology. *Nature* 102: 88–89.

Pear, T. H. and Smith, G. E. (1917). *Shell-shock and Its Lessons*. Manchester: Manchester University Press.

Pennebaker, J. W. (1993). Putting stress into words: Health, linguistic, and therapeutic implications. *Behaviour Research and Therapy* 31: 539–548.

Pennebaker, J. W. (1995). *Emotion, Disclosure and Health*. Washington, DC: American Psychological Association.

Pennebaker, J. W. (1997). Writing about emotional experiences as a therapeutic process. *Psychological Science* 8: 162–166.

Pihkala, P. (2020). Anxiety and the ecological crisis: An analysis of eco-anxiety and climate anxiety. *Sustainability* 12: 7836.

Ray, S. J. (2020). *A Field Guide to Climate Anxiety*. Oakland: University of California Press.

Reser, J. P. et al. (2012). Public risk perceptions, understandings and responses to climate change and natural disasters in Australia 2010 and 2011. NCCARF: Griffith University.

Richell, R. A. et al. (2003). Theory of mind and psychopathy: Can psychopathic individuals read the language of the eyes. *Neuropsychologia* 41: 523–526.

Rivers, W. H. R. (1918). The repression of war experience. *The Lancet* 2: 173–177.

Rude, S. S. and Haner, M. L. (2018). Individual differences matter. *Clinical Psychology: Science and Practice* 25: 12230.

Rydell, R. J. and McConnell, A. R. (2006). Understanding implicit and explicit attitude change: A systems of reasoning analysis. *Journal of Personality and Social Psychology* 91: 995–1008.

Sai, L. et al. (2021). Theory of mind, executive function, and lying in children: A meta-analysis. *Developmental Science* 24(5): e13096.

Sangervo, J. et al. (2022). Climate anxiety: Conceptual considerations and connections with climate hope and action. *Global Environmental Change* 76: 102569.

Sarason, I. G. and Sarason, B. R. (1990). Test anxiety. In *Handbook of Social and Evaluation Anxiety* (pp. 475–495). Boston, MA: Springer US.

Scheier, M. F. et al. (1989). Dispositional optimism and recovery from coronary artery bypass surgery: The beneficial effects on physical and psychological well-being. *Journal of Personality and Social Psychology* 57: 1024–1040.

Schulz, R. et al. (1996). Pessimism, age and cancer mortality. *Psychology and Aging* 11: 304–309.

Schwartz, S. E. et al. (2022). Climate change anxiety and mental health: Environmental activism as buffer. *Current Psychology* 42: 16708–16721.

Searle, K. and Gow, K. (2010). Do concerns about climate change lead to distress? *International Journal of Climate Change Strategies and Management* 2: 362–379.

Segerstrom, S. S., Taylor, S. E., Kemeny, M. E., and Fahey, J. L. (1998). Optimism is associated with mood, coping and immune change in response to stress. *Journal of Personality and Social Psychology* 74: 1646–1655.

Seligman, M. E. (1972). Learned helplessness. *Annual Review of Medicine* 23: 407–412.

Seligman, M. E. et al. (1979). Depressive attributional style. *Journal of Abnormal Psychology* 88: 242–247.

Seligman, M. (2002). *Authentic Happiness. Using the New Positive Psychology to Realize Your Potential for Lasting Fulfilment*. New York: Free Press.

Sharot, T. (2012). *The Optimism Bias. Why We're Wired to Look on the Bright Side*. London: Robinson Books.

Shi, J. et al. (2016). Public perception of climate change. *Nature Climate Change* 6: 759–762.

Slovic, P. et al. (2002). Rational actors or rational fools: Implications of the affect heuristic for behavioral economics. *The Journal of Socio-Economics* 31: 329–342.

Spielberger, C. D. (Ed.) (1966). *Anxiety and Behavior*. London: Academic Press.

Stevens, D. et al. (2001). Recognition of emotion in facial expressions and vocal tones in children with psychopathic tendencies. *The Journal of Genetic Psychology* 162: 201–211.

Talwar, V. and Lee, K. (2008). Social and cognitive correlates of children's lying behaviour. *Child Development* 79: 866–881.

Tam, K. P., Chan, H. W. and Clayton, S. (2023). Climate change anxiety in China, India, Japan, and the United States. *Journal of Environmental Psychology* 87: 101991.

Taylor, S. (2020). Anxiety disorders, climate change, and the challenges ahead: Introduction to the special issue. *Journal of Anxiety Disorders* 76: 102313.

Tsakiridou, E. et al. (2008). Attitudes and behaviour towards organic products: An exploratory study. *International Journal of Retail and Distribution Management* 36: 158–175.

Tschakert, P., Tutu, R. and Alcaro, A. (2013). Embodied experiences of environmental and climatic changes in landscapes of everyday life in Ghana. *Emotion, Space and Society* 7: 13–25.

Twenge, J. M. and Campbell, K. W. (2009). *The Narcissism Epidemic. Living in the Age of Entitlement*. New York: Atria.

UK Climate Change Risk Assessment 2017 Evidence Report (2017). https://www.theccc.org.uk/uk-climate-change-risk-assessment-2017/synthesis-report/. Accessed 19 November 2024.

Unilever (2013). Unilever Sustainable Living Plan. https://assets.unilever.com/files/92ui5egz/production/910902bc7c415bbb6fdd0d9474ccb10da1e7c671.pdf/slp_unilever-sustainable-living-plan-2013.pdf. Accessed 11 July 2023.

United Nations Office for Disaster Risk Reduction (UNDRR) Launch of Global Assessment Report on Disaster Risk Reduction (GAR2019). https://www.undrr.org/

Upham, P., Dendler, L. and Bleda, M. (2011). Carbon labelling of grocery products: Public perceptions and potential emissions reductions. *Journal of Cleaner Production* 19: 348–355.

van der Linden, S. (2015). The social-psychological determinants of climate change risk perceptions: Towards a comprehensive model. *Journal of Environmental Psychology* 41: 112–124.

van der Linden, S. (2017). Determinants and measurement of climate change risk perception, worry, and concern. *The Oxford Encyclopaedia of Climate Change Communication*. Oxford: Oxford University Press.

van Valkengoed, A. M. (2023). Climate anxiety is not a mental health problem. But we should still treat it as one. *Bulletin of the Atomic Scientists* 79: 385–387.

Verlie, B. (2022). *Learning to Live with Climate Change: From Anxiety to Transformation*. London: Routledge.

Verplanken, B. and Roy, D. (2013). My worries are rational, climate change is not. *PloS One* 8: e74708.

Verplanken, B. et al. (2020). On the nature of eco-anxiety: How constructive or unconstructive is habitual worry about global warming. *Journal of Environmental Psychology* 72: 101528.

Whitmarsh, L. et al. (2022). Climate Anxiety: What predicts it and how is it related to climate action? *Journal of Environmental Psychology* 83: 101866.

Williams, B. et al. (1999). Development of inhibitory control across the lifespan. *Developmental Psychology* 35: 205–213.

Willis, J. and Todorov, A. (2006). First impressions: Making up your mind after a 100-ms exposure to a face. *Psychological Science* 17: 592–598.

World Health Organization (2017). Climate change and human health. http://www.who.int/globalchange/en/. Accessed 15 March 2017.

World Resource Institute (2014). www.wri.org/blog/2014/05/history-carbon-dioxide-emissions. Accessed 10 July 2016.

Wray, B. (2023). *Generation Dread: Finding Purpose in an Age of Climate Crisis*. Canada: Penguin.

Wu, J., Snell, G. and Samji, H. (2020). Climate anxiety in young people: A call to action. *The Lancet Planetary Health* 4: e435–e436.

Yale Climate Opinion Maps (2023). https://climatecommunication.yale.edu/visualizations-data/ycom-us/. Accessed 19 November 2024.

Yang, J. Z. and Kahlor, L. A. (2012). What, me worry? The role of affect in information seeking and avoidance. *Science Communication* 35: 189–212.

Yealland, L. R. (1918). *Hysterical Disorders of Warfare.* London: Macmillan

Zajonc, R. (1980). Feeling and thinking: Preferences need no inferences. *American Psychologist* 35: 151–175.

Zeidner, M. and Matthews, G. (2010). *Anxiety 101.* London: Springer.

Zhao, N. and Zhou, G. (2020). Social media use and mental health during the COVID-19 pandemic: Moderator role of disaster stressor and mediator role of negative affect. *Applied Psychology: Health and Well-Being* 12: 1019–1038.

For Product Safety Concerns and Information please contact our EU
representative GPSR@taylorandfrancis.com
Taylor & Francis Verlag GmbH, Kaufingerstraße 24, 80331 München, Germany